Laser Processing in Manufacturing

Engineering Aspects of Lasers Series

Series editor
Dr T.A. Hall, *Reader in Physics, University of Essex*

In the late 1960s and early 1970s the laser was still something of a scientific curiosity with only a limited practical use. The extent of the four volumes in this series shows the enormous change that has happened since that time. The laser is now an indispensable addition to the toolbox of the engineer and scientist. The progress from the time when the laser was often dubbed 'a solution in search of a problem' to today, when engineers of all disciplines frequently use lasers as a matter of course, is a remarkable transformation. Even so, the use of lasers in engineering and other walks of life is still in its infancy and has been held back partly by their relatively high cost and in some cases by their inconvenience in use. The cost of many laser systems has been falling for several years, they are becoming much more convenient to use and no longer need trained personnel to operate them. As these problems are overcome, lasers will find wider and wider application and there is an ever increasing need for engineers and scientists, who perhaps have little interest in lasers themselves, to have access to an authoritative source which not only acts as an introduction but also takes the reader up to the latest developments in laser applications.

The four books of the series 'Engineering Aspects of Lasers' arose from a series of laser workshop courses which have been held each year at the University of Essex since 1979. These courses have evolved very considerably since their inception but aspects of their organization have not changed – the contents of the courses have always been co-ordinated by the recognized international authority in each subject area and the lectures given by experts in the particular field from industry, government laboratories or universities. When the idea of publishing a series of books based upon the contents of these courses was first suggested the course co-ordinators at that time quite naturally became the editors of each volume and the lecturers were asked to contribute.

The workshops are self-supporting courses which originally formed part of the MSc degree in Lasers and their Applications. There are many people who have contributed much to these courses over the years and have made them the success that they are. I would like to express my gratitude to them all. The courses and the MSc were the brainchild of T.P. Hughes who was then Reader in Physics at Essex University. Without his foresight, hard work and determination in setting up the courses, this series of books would not have been written.

Other titles in this series

Optical Methods in Engineering Metrology
Edited by D.C. Williams

Nonlinear Optics in Signal Processing
Edited by R.W. Eason and A. Miller

Advances in Optical Communications
Edited by N. Doran and I. Garrett

Laser Processing in Manufacturing

Edited by

R.C. Crafer

Partner
Abington Consultants

and

P.J. Oakley

Consultant

CHAPMAN & HALL
London · Glasgow · New York · Tokyo · Melbourne · Madras

Published by Chapman & Hall, 2–6 Boundary Row, London SE1 8HN

Chapman & Hall, 2–6 Boundary Row, London SE1 8HN, UK

Blackie Academic & Professional, Wester Cleddens Road, Bishopbriggs, Glasgow G64 2NZ, UK

Chapman & Hall Inc., 29 West 35th Street, New York NY10001, USA

Chapman & Hall Japan, Thomson Publishing Japan, Hirakawacho Nemoto Building, 6F, 1-7-11 Hirakawa-cho, Chiyoda-ku, Tokyo 102, Japan

Chapman & Hall Australia, Thomas Nelson Australia, 102 Dodds Street, South Melbourne, Victoria 3205, Australia

Chapman & Hall India, R. Seshadri, 32 Second Main Road, CIT East, Madras 600 035, India

First edition 1993

Typeset in 10/12 pt Times by Graphicraft Typesetters Ltd, Hong Kong
Printed in Great Britain by the University Press, Cambridge

ISBN 0 412 41520 8

A catalogue record for this book is available from the British Library

Library of Congress Cataloging-in-Publication data available.

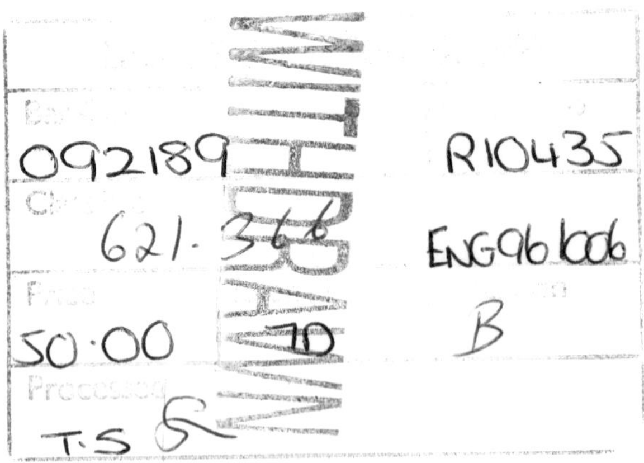

Contents

Contributors

M. Adams
Partner, Beech House Associates
(formerly Technical Director,
Electrox Limited)

R.C. Crafer
Partner, Abington Consultants
(formerly, The Laser Centre, TWI)

M.C. Gower
Exitech Limited
Long Hanborough
Oxford

D. Greening
V&S Scientific (London) Limited
Potters Bar
Hertfordshire

A. May
Laser Lines Limited
Banbury
Oxfordshire

P.J. Oakley
Consultant
(formerly, The Laser Centre, TWI)

T.M.W. Weedon
Lumonics Limited
Rugby
Warwickshire

C. Williams
Sales Engineer
Laser Ecosse Limited
Dundee
Angus

Preface

Laser processing of material, to effect a change in shape or form, has been available for thirty years. The use of lasers was not widespread initially, but did fire the imagination of the popular press and the general public. More recently, the situation has reversed, with laser processing being used as part of the production route for many items, usually without the purchaser realizing it. This lower profile and wider use has allowed the laser to become 'just another machine tool' to many production engineers. However, because of the uniqueness of the laser's capabilities and processing route, increased usage does require the education and training of staff at different levels in order to build up the necessary infrastructure in industry. This is where our book is targeted, and hopefully it will play its part in this education and training process.

The origins of the book lie in the Laser Technology MSc course run at Essex University each year from the late 1970s until 1990. In the spring term of the course, a series of one-week laser applications workshops were organized by people active in the respective fields, and were open to industrial delegates as well as the MSc students. The 'Applications of Lasers in Welding, Cutting, and Surface Treatment' was one of these intensive workshops, which called on a range of experts from the UK laser fraternity to impart their knowledge in both formal and informal sessions. The book is based on the formal presentations made by contributors to the course. The MSc has been superseded by more specific courses. However, at the time of writing the 'Applications of Lasers in Welding, Cutting, and Surface Treatment' workshop is continuing as a stand-alone event for industrial delegates.

The book is based on refined versions of the formal presentations made as part of the course. Both carbon dioxide and solid-state lasers have been part of the course since its inception, and principles, engineering, and usage are covered. Excimer laser engineering and applications were introduced more recently to the course as a result of their growing significance to laser materials processing. The excimer laser and its actual and potential applications form an extensive part of the book, because we believe many people are totally unfamiliar with this area of our subject. With such a range of authors, some repetition was unavoidable. This has been eliminated where it was a major concern, but left where the overlap was minor, or the authors brought a different perspective to the topic of discussion. Similarly, rather than impose a uniform writing style on the authors, the range of styles has been left largely as originated.

In preparing this book, the editors would like to acknowledge and thank several groups of people. Firstly there are the staff of the Physics Department of Essex University, who started and continue to run the course on which the book is based. Particular mention must be made of Carol Snape and David Lovett, who have ensured the smooth running of the course despite gales and snowstorms. Secondly, we must thank the authors, not only for their book chapters, but also for their continuing contributions to the course. Lastly, we would also like to thank the MSc students and industrial participants of the courses who, generally after a quiet start to the week, have soon overcome their reserve and played their part in making the courses lively, stimulating events, often challenging the speakers with inventive questioning.

Roger Crafer
Peter Oakley

1

An introduction to lasers

R.C. Crafer

1.1 DEFINITION

The internationally recognized term LASER is an acronym standing for Light Amplification by Stimulated Emission of Radiation. The word laser can be either a noun or an adjective. As a noun it represents an item of equipment utilizing the principle above to produce a light beam having remarkable properties. In this usage it is often preceded by the name of the principal substance involved in its operation, for example Neodymium: YAG, Carbon Dioxide, Excimer, etc. As an adjective it generally describes the use to which a laser (noun) can be put. Examples here would include materials processing (e.g. welding), ophthalmology, range finding, communications, etc. This chapter introduces the physical ideas involved and some of the concepts and jargon to be encountered in later chapters.

1.2 BRIEF HISTORY

The laser as an identifiable item of equipment made its first appearance with the work of Maiman (1960). However a number of key advances during the preceding century contributed to its development. Some familiarity with these will be helpful in understanding the laser and its remarkable properties.

In the mid-nineteenth century the behaviour of light was quite successfully explained by the laws of Ampère, Gauss and Lenz, which are better known today in combined form as Maxwell's Equations (see Chapter 2, section 2.10.1). These were, and still are, highly successful in explaining large-scale or macroscopic behaviour, but fail to explain interactions on an atomic scale. In particular, these equations predicted the occurrence of a so-called ultraviolet catastrophe which fortunately was not observed in practice. This anomaly between theory and experiment was eventually resolved by the German physicist Planck, who postulated a particulate or quantum

Laser Processing in Manufacturing. Edited by R.C. Crafer and P.J. Oakley.
Published in 1993 by Chapman & Hall, London. ISBN 0 412 41520 8

nature of light itself. This simple but elegant approach removed the ultraviolet catastrophe and brought prediction and observation once more into agreement. In 1913, the Danish physicist Bohr extended Planck's ideas to atomic energy structure and was able to explain the absorption and emission spectra of atoms in terms of transitions beween atomic energy levels. This was followed by Einstein's (1917) pivotal paper on the interaction between atomic energy levels and light, which led to the prediction of a hitherto unobserved effect. This effect, now called Stimulated Emission, is central to the operation of all laser devices. In the early 1930s, following a decade in which the empirical work of Planck and Bohr was put on to a sound theoretical footing by workers such as de Broglie and Schrödinger, stimulated emission was actually observed in a gas discharge by the German physicist Ladenburg. However the amazing potential of Stimulated Emission was not at that time recognized.

The first device utilizing stimulated emission was reported by Gordon, Zeiger and Townes (1955). This device achieved Microwave Amplification by Stimulated Emission of Radiation (hence the acronym MASER). Following several years of intense debate concerning the possibility of this microwave effect occurring at optical or visible wavelengths, the matter was finally put to rest in 1960 with the invention of the ruby laser by Maiman of the Hughes Research Laboratories. Once the principle had been established, that the optical maser or laser was in fact a possible device, there followed a decade of intense activity during which most of the laser types we know today were established. In particular, and in chronological order, the Helium Neon laser was investigated and developed by Javan *et al.* (1961), Neodymium:YAG by Geusic *et al.* (1964), and Carbon Dioxide by Patel (1964). The only major industrial type to be developed outside that decade was the excimer laser in 1976.

1.3 BACKGROUND PHYSICS

1.3.1 Atoms, ions and molecules

Matter is composed of atoms. Each atom consists of a nucleus surrounded by one or more electrons. The nucleus is incredibly small, typically one hundredth of a millionth of a millionth of a metre (10^{-14} m) across. It consists of one or more protons and a number of neutrons. Protons are positively charged particles, neutrons have no charge at all. Electrons are negatively charged and are typically one thousandth of a millionth of a metre (10^{-9} m) or 1 nanometre across. Atoms themselves are electrically neutral, and have equal numbers of protons and electrons. The number of protons is called the atomic number and is designated by the letter Z. Under certain circumstances, atoms may gain or lose one or more electrons; such charged atoms are called ions. Ions differ in spectroscopic and chemical properties from their neutral counterparts. Atoms cluster together in stable groups called molecules and crystals. In some cases the bond between atoms is caused by exchange of electrons, and in other cases by sharing of electrons. The former bonding is called ionic, and the latter covalent.

1.3.2 Energy levels

Atomic systems exist only at well-defined energies. For atoms and ions the energies of interest are due to the electrons farthest from the nucleus. It is instructive to consider an atom as consisting of a number of electrons orbiting around a nucleus. Electrons orbiting near to the nucleus are bound strongly since the Coulomb force law is obeyed. The outermost electrons are bound only weakly. Quantum mechanics imposes a constraint that only specific orbital radii are permitted. As energy is added to the atom, the outermost electron assumes a progressively larger orbit until, finally, sufficient energy is imparted to remove it completely from the nucleus, whereupon the atom becomes an ion. Corresponding to each allowed orbit of the outermost electron, there is a definite energy level.

Molecules also exhibit energy levels due to electrons, but in addition, each atom can vibrate with respect to the others or rotate about a number of common axes. Both vibration and rotation give rise to energy levels.

1.3.3 Light

Light is a form of electromagnetic radiation in and around the visible region of the spectrum. For the purposes of practical laser interest, it spans the range from the ultraviolet at 190 nm wavelength to the far infrared at 1 mm wavelength.

Light may be treated either as a classical wave motion using Maxwell's equations, or as a quantum mechanical stream of particles called photons. Generation and amplification in a laser medium is best treated by quantum mechanical methods. Propagation within an amplifying medium and its transmission from the laser to the workpiece is described adequately by Maxwell's equations.

Classically, light is described as an oscillating electric field with an associated magnetic field. Both fields vary sinusoidally with the same frequency and propagate in the same direction, but are polarized at right angles to each other as shown in Fig. 1.1. Both fields alternate ν times every second so ν is known as the frequency of the light. The distance between successive waves is the wavelength and is designated by λ. The wave itself propagates through space with a velocity u given by the product of the frequency ν and the wavelength λ, i.e. $u = \nu \times \lambda$.

Quantum mechanically, light is a stream of photons, each with energy E given by the product of the frequency ν and Planck's constant h, i.e.

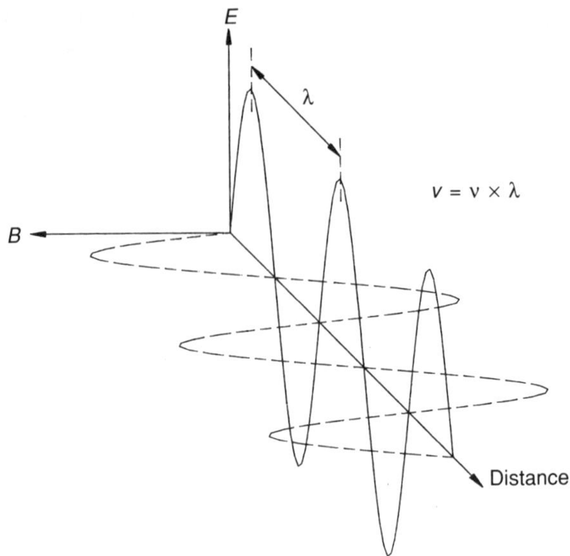

Fig. 1.1 Propagation of light.

$E = h \times \nu$, where h is a universal constant of approximate value 6.6×10^{-34} joule seconds.

1.3.4 Light matter interactions

There are two relevant interactions. The first of these causes an atom, ion or molecule to jump from one energy level to another with the associated emission or absorption of a photon. The second is a direct interaction with free electrons either in a plasma or within the conduction band of solid materials.

Light interacts with atoms, ions or molecules only when the photon energy E is very near to the difference of two energy levels. If the level energies are E_1 and E_2, where E_2 exceeds E_1, only photons with energies very near to $E_2 - E_1$ can be emitted or absorbed. There are three distinct possibilities:

1. Level 2 relaxes spontaneously to level 1 emitting a photon randomly in time, phase and direction. The probability of this happening is given by the coefficient A_{21}. (The suffix $_{21}$ means *from* level 2 *to* level 1).
2. A photon is absorbed causing a jump from level 1 to level 2. The probability of this occurring is given by the coefficient B_{12}.
3. An incident photon stimulates the emission of a second photon causing a jump from level 2 back to level 1. Both stimulated and stimulating

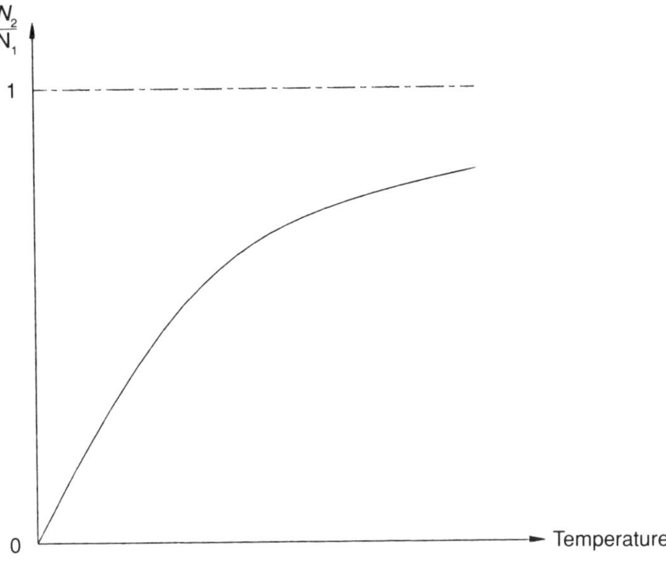

Fig. 1.2 Thermal level population.

photons are identical and indistinguishable. The probability of this happening is given by the coefficient B_{21}.

The coefficients A and B are called Einstein coefficients. For any given pair of simple levels B_{21} always equals B_{12}, hence if level populations are N_1 and N_2 respectively, the net absorption depends on $N_1 - N_2$. There is also an important relationship between A_{21} and B_{21}, namely $A_{21}/B_{21} = 8\pi h\nu^3/c^3$, where c is the velocity of light. Spontaneous emission (the A coefficient) is thus the dominant energy loss mechanism at high frequencies (short wavelengths), whereas stimulated emission (the B coefficient) can dominate at low frequencies (long wavelengths).

1.3.5 Normal population

In reality, energy level structures can be extremely complicated, but we will simplify matters by considering just a single pair of level L_1 and L_2. The ratio of the numbers populating these levels is obtained from the well-known Maxwell-Boltzmann distribution and is given by

$$N_2/N_1 \approx \exp(-(E_2 - E_1)/kT) \tag{1.1}$$

where k is Boltzmann's constant (approximately 1.4×10^{-23} joules per kelvin) and T is the kelvin temperature. A plot of equation (1.1) is shown in Fig. 1.2. At low temperatures, only the lower level is significantly occupied. As the temperature rises however, the higher level population increases until, at the highest temperatures, both levels are significantly occupied. One underlying feature is that no matter which pair of levels is chosen, the population of the lower level always exceeds that of the higher. This is called a normal population.

1.3.6 Inverted population and amplification

Normal populations, i.e. those where N_1 exceeds N_2, occur naturally and result in a net loss of light through absorption exceeding stimulated emission. However, if the reverse situation could be engineered, such that N_2 exceeded N_1, then light would be amplified and we would have Light Amplification by Stimulated Emission of Radiation, or in other words a LASER. This reversal of the normal population, or 'population inversion' as it is called, is a prerequisite of all laser devices.

A normal population results from thermodynamic equilibrium conditions. If equilibrium is disturbed, a population inversion can occur as the system reacts to the disturbance and returns to equilibrium. This is the method successfully applied in the ruby laser.

Alternatively, two different systems can be brought together in such a way that equilibrium is upset for the duration of the interaction. This second method is utilized in the majority of gas lasers, e.g. Helium, Neon and Carbon Dioxide, and also in solid-state lasers such as Neodymium: YAG.

1.3.7 Achieving inversion

Many successful continuous lasers operate on the so-called, four-level principle. The lowest and highest energy levels labelled L_1 and L_2 in Fig. 1.3 are chosen for their effectiveness in utilizing some external pumping process to produce a large population in level L_2, although this is not a population inversion between L_2 and L_1. Most pumping processes operate by colliding particles (electrons or photons) with level L_1. The pumping rate is proportional to the product of particle numbers and L_1 population. This population N_1 should thus be as large as possible and hence the ground level is often chosen as the lower of the two pumping levels. Similarly, to maintain a large excited population in L_2, its lifetime should be very long to minimize energy losses. Two intermediate levels L_3 and L_4 are chosen for their ability to interact strongly with external light beams. These are the lasing levels. Both spontaneous and stimulated emission compete to depopulate L_3, so its lifetime must also be long. Conversely, any population in L_4 must be removed rapidly to maximize inversion, hence its lifetime must be short.

The main reason for separating the pumping levels 1 and 2 from the

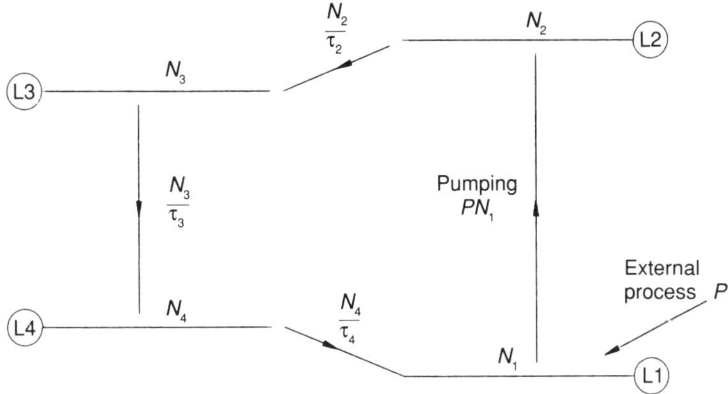

Fig. 1.3 Four-level energy scheme.

lasing levels 3 and 4, is their contradictory requirements. Effective pumping requires a large population in the lower level while effective lasing demands a vanishingly small population. Likewise, effective lasing demands significant radiative decay from the upper level whereas effective pumping requires a minimum of decay. A successful laser requires two suitable pairs of levels to be coupled in such a way that energy flows freely between them.

1.3.8 Rate equations

On a more formal basis, suppose we let the energy flows be as shown in Fig. 1.3. Decay is only to the level immediately beneath. The pumping or energy input rate is written as PN_1, where P is the strength of the external pumping process such as electric discharge excitation or exposure to flash lamp light. Spontaneous decay rates from levels L_2, L_3 and L_4 are N_2/τ_2, N_3/τ_3 and N_4/τ_4 respectively. It is also assumed that the temperature is low so that thermal excitation into L_4 can be neglected and that no external light is present so that stimulated emission from L_3 to L_4 may also be neglected.

In the steady state, the pumping rate must equal the decay rates. Also the total population of the pumping levels must be constant – let us call this N. Combining these requirements yields

$$N_2 - N_3 = PN(\tau_3 - \tau_4) \qquad (1.2)$$

The population inversion is thus seen to be proportional to the strength of the external process, the total population N of the pumping levels, and the difference between the lifetimes of the upper and lower laser levels L_3 and L_4.

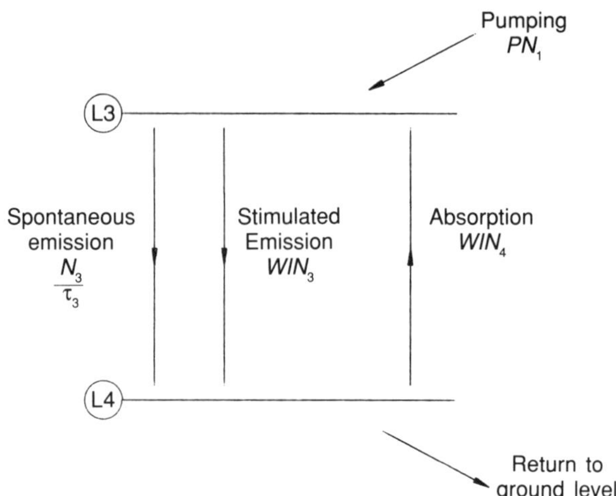

Fig. 1.4 Two-level amplifier scheme.

1.3.9 The two-level amplifier

The simplest operational laser device is the two-level, travelling-wave amplifier, consisting of a volume of transparent material supporting a population inversion. Light propagating through the material acquires energy via stimulated emission and emerges amplified. Consider the lasing levels of our four-level model. Other levels may be disregarded provided that energy loss processes are negligible (Fig. 1.4).

External pumping into L_3 is still PN. Incoming light perturbs the situation by adding two more energy flows – stimulated emission out of L_3 and absorption out of L_4. These have magnitude WIN_3 and WIN_4, where I is the intensity of the light beam and W is a constant taking account of all other physical parameters.

In the steady state

$$N_3 - N_4 = PN(\tau_3 - \tau_4)/(1 - I/I_s) \qquad (1.3)$$

where I is a constant of the amplifier called the saturation parameter. For $I < I_s$, the population inversion tends towards the value given by equation (1.2); this is called the small signal condition.

Now, consider this same interaction from the viewpoint of the light beam, as depicted in Fig. 1.5. In travelling a small distance δx, the light encounters a total population inversion of $(N_3 - N_4)\delta x$. The extra intensity δI acquired through stimulated emission is therefore $WI(N_3 - N_4)\delta x$. Solving for the logarithmic gain G we obtain

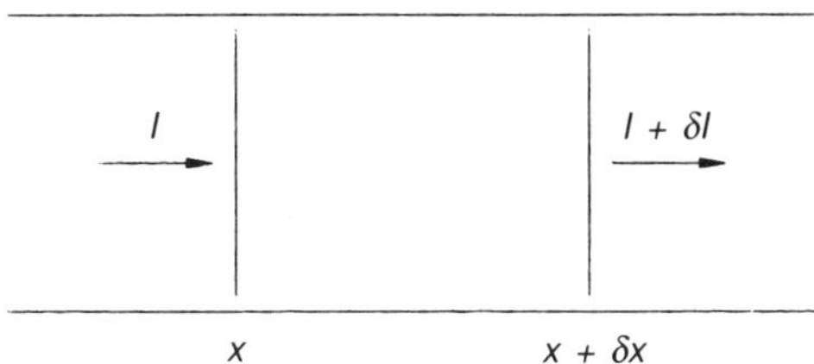

Fig. 1.5 Uptake of optical energy.

$$G = G_0/(1 + I/I_s) \tag{1.4}$$

where $$G_0 = WPN\ (\tau_3 - \tau_4)$$

G_0 is called the small signal gain. In this simple model it varies directly with the strength of the pumping process, the total population of the pumping levels, and the difference of the lasing level lifetimes.

1.4 LASER FUNDAMENTALS

1.4.1 Resonators and cavities

The words resonator and cavity derive from the different backgrounds of the original laser investigators. Cavity refers to an electrically resonant enclosure. Resonator refers to a partially closed acoustic body that will support a musical note (e.g. a Helmholtz resonator).

1.4.2 Travelling-wave resonators

Most people are familiar with audio amplifiers using transistor devices. These amplifiers are unidirectional in that the input signal is always applied to one terminal, and the output taken from another. Feedback is taken from the output, via an external circuit, to the input. The signal circulates around this loop in one direction only; if the feedback is such as to augment the signal, then oscillation occurs. The optical analogue is the travelling-wave resonator, in which the optical signal circulates round the device in the same direction using a system of optical elements. This type of resonator is very specialized, and seldom if ever used in industry.

1.4.3 Standing-wave resonators

The more common type of resonator makes use of the fact that laser amplifiers will amplify light travelling in any direction. The feedback path is identical to the amplifying path with the exception that the direction of

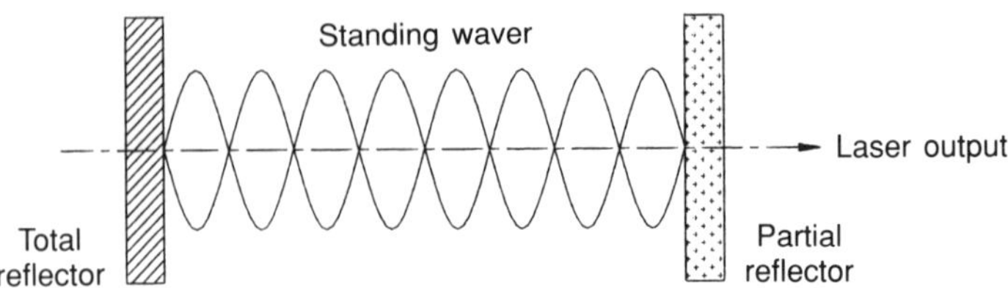

Fig. 1.6 Simple optical resonator.

propagation is reversed. Resonators such as these are called standing-wave resonators. The simplest such resonator consists of two plane mirrors facing each other along a common axis, as shown in Fig. 1.6.

If we combine two waves travelling in opposite directions along the resonator axis, stable resonance patterns or standing waves occur whenever the separation of the mirrors is an integral number of half wavelengths, or

$$L = n\lambda/2 \tag{1.5}$$

Each resonance characterized by a different value of n is called a mode of the resonator. In the case described by equation (1.5), the variation is purely axial, and the modes are known as axial modes. Such resonators are seldom used, for two reasons:

1. Light beams expand as they propagate and after several traverses between plane mirrors would spill over the edges of the mirrors, causing large losses and low efficiencies.
2. The alignment of the mirrors is highly critical.

To circumvent these disadvantages, most practical resonators have curved mirrors. In a simple case, both mirrors are sufficiently concave to counteract natural diffraction spreading. This also relaxes mirror alignment since for small offsets there is always one point on each mirror aligned with the other. A side effect of mirror curvature is that the number of standing waves depends on the exact path traversed, so that, in addition to the axial modes, there are also off-axis or transverse modes. These are usually described as Transverse Electromagnetic or TEM for short, followed by two suffix digits and an optional asterisk. The two digits refer to the degree of mode structure, thus TEM_{00} is purely axial, TEM_{10} has a simple two-spot structure in the x direction, and in general TEM_{n0} has an $n + 1$ spot structure. Likewise the second digit refers to structure in the y direction. Both x and y structures can be combined, as for example TEM_{32}. Figure 1.7 shows examples of these modes. The $x - y$ type description is relevant to lasers where optical asymmetries define an x direction. In many lasers such asymmetries are weak, and the modes have a circular symmetry. Here the two digits are followed by an asterisk (enunciated star). As an

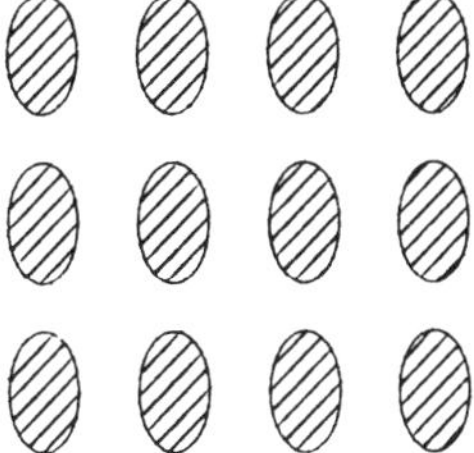

Fig. 1.7 TEM_{32} mode.

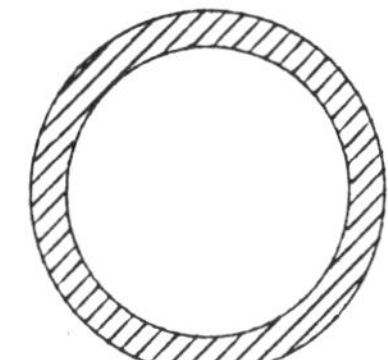

Fig. 1.8 TEM_{01*} mode.

example, TEM_{01*} consists of a hollow annular ring centred on the axis, as shown in Fig. 1.8. It is not unusual for lasers to operate in several $x - y$ or * modes at once, and in this context the phrase 'combination of low-order transverse modes' is often encountered.

1.4.4 Mode stability

Mode stability within a curved mirror resonator was thoroughly investigated early in the course of laser development. A concise and useful description is the stability diagram of Fox and Li (1961) on which any combination of mirror curvatures and separations can be plotted as a single point, (Fig. 1.9).

Hyperbolic curves represent the limits of stability. In the shaded region between these curves and the axes, light is refocused periodically by the mirrors and the volume occupied, the so-called mode volume, remains fixed and well defined. Resonators lying within these areas are called stable resonators.

Resonators lying beyond the curves, or in the other quadrants, are called unstable resonators. These have their own, totally different type of mode structure in which successive reflections never retrace their previous paths. An example of an unstable resonator configuration is shown in Fig. 1.10.

Neodymium:YAG lasers use stable resonators almost exclusively. CO_2 lasers use both types. Stable resonators are ideal for cases where light is extracted through one partially transmitting mirror. Very high power CO_2

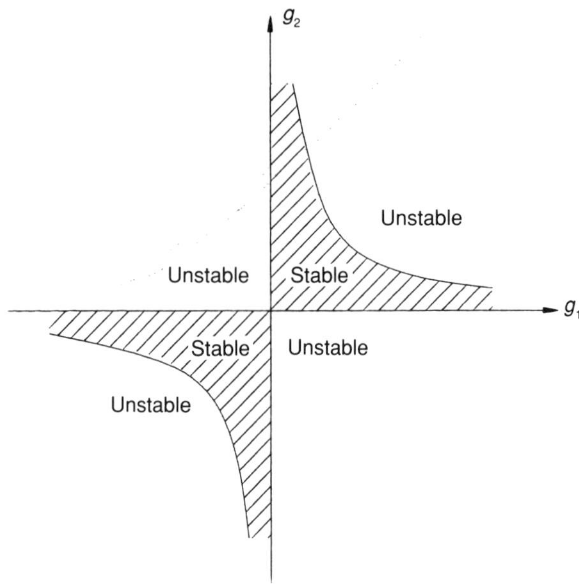

Fig. 1.9 Fox and Li stability diagram.

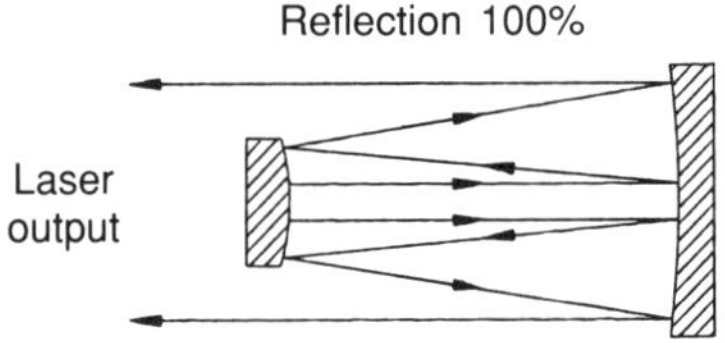

Fig. 1.10 Unstable resonator.

lasers cannot use this approach for various reasons, and therefore employ unstable resonators.

1.4.5 Oscillators

Self-excited oscillators consist of a laser amplifier enclosed in the positive feedback loop of a resonator. Threshold reflectivity is the minimum reflectivity necessary to achieve laser output. It is that reflectivity which makes the round trip losses equal to the small signal gain. Immediately above the threshold reflectivity the output power rises. As the reflectivity approaches unity, the output power falls to zero. Between these two extremes, the output power passes through a maximum value at the optimum reflectivity.

1.4.6 Fundamental mode

Reference is often made to a fundamental or ideal laser mode. This is the TEM_{00}, or so-called Gaussian mode. Many industrial lasers strive to achieve it, but few succeed, particularly at high powers. The fundamental mode is of interest since the beam from a fundamental-mode laser retains its size longer than for any other mode and can be focused to a finer and more intense spot. The intensity I of a Gaussian beam is related to its peak intensity I_0 by the relationship:

$$I = I_0 \exp -(r/r_0)^2 \tag{1.6}$$

where r is the distance from the beam axis and r_0 is a length called the beam (or spot) radius. In physical terms, the intensity at $r = r_0$ is reduced by a factor e^2 ($\approx$7.3) from its peak intensity. The term **spot size**, often quoted in the literature, usually but not always refers to beam diameter, i.e. $2r_0$.

1.4.7 Beam quality and divergence

All laser beams diverge. That is, they become larger as they travel. A typical divergence for a CO_2 laser is 2 milliradians (mrad), which means that for every metre the beam travels, it expands its diameter by 2 mm. The actual divergence angle depends on the laser wavelength, beam diameter and mode structure. A more useful quantity is the product of beam diameter as it leaves the laser and divergence angle a long distance away. The ratio of this actual product for any particular beam to its value for a gaussian beam is a useful quality factor for the laser.

1.5 THERMODYNAMICS OF LASER OPERATION

1.5.1 Introduction

Lasers are basically power transformers. Power supplied in various forms, predominantly electrical, chemical or optical, is converted to light with heat as the main by-product. The transformation process is very inefficient, so that most of the power is converted to heat. Since most lasers have some limiting values of temperature (or thermal gradient) above which they cease to function, removal of this unwanted heat is critical. In the design of industrial lasers, the useful power output depends on this ability to remove heat.

1.5.2 Cylindrical geometry

In solid lasers (e.g. Ruby, Nd:YAG), heat flows through the medium by conduction. Certain low power CO_2 lasers can also be analysed on this

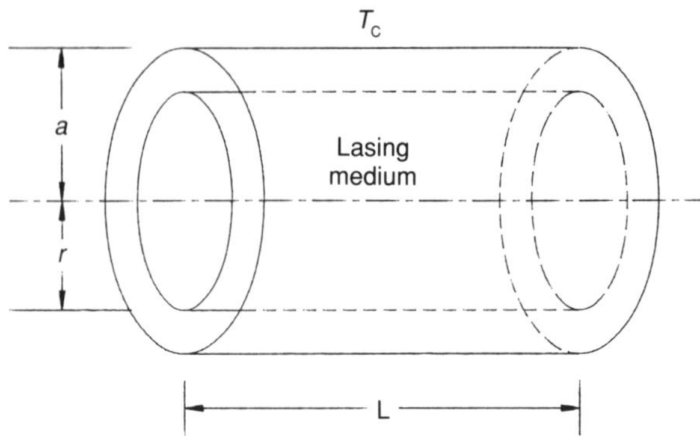

Fig. 1.11 Conduction model of laser heat balance.

conduction-like basis. In the latter case, fluid velocity due to natural or forced convection must be slow compared with thermal transport due to diffusion. Consider a laser as a cylinder of radius a and length L, cooled by an external coolant at temperature T_c; see Fig. 1.11. Assume waste heat is deposited uniformly within the cylinder at a rate Q per unit volume. Applying Fourier's law of heat conduction to an axial cylinder of radius $r < a$, we obtain:

$$\text{Heat generated within cylinder} = \pi r^2 Q \text{ per unit length}$$
$$\text{Heat loss from cylinder} = -2K\pi r \mathrm{d}T/\mathrm{d}r \text{ per unit length}$$

where T is the temperature and K the thermal conductivity.

In a continuous-wave laser these rates must balance, resulting in the differential equation

$$\mathrm{d}T/\mathrm{d}r = -Qr/2K \tag{1.7}$$

for which the appropriate solution is

$$T = (Qa^2/4K)\,(1 - r^2/a^2) + T_c \tag{1.8}$$

The temperature T is highest along the axis, where it has the value

$$T_{\text{axis}} = Qa^2/4K + T_c \tag{1.9}$$

Equating this to the limiting temperature T_{lim}, and replacing Q by

$$Q = P/\pi\xi a^2 L \tag{1.10}$$

where P is the laser output power, and ξ the transformation efficiency assumed <<1, we obtain

$$P = 4\pi\xi KL(T_{\text{lim}} - T_c) \tag{1.11}$$

This is a most important result. It states that the only dimension affecting the laser output power is length, and that diameter does not enter into the

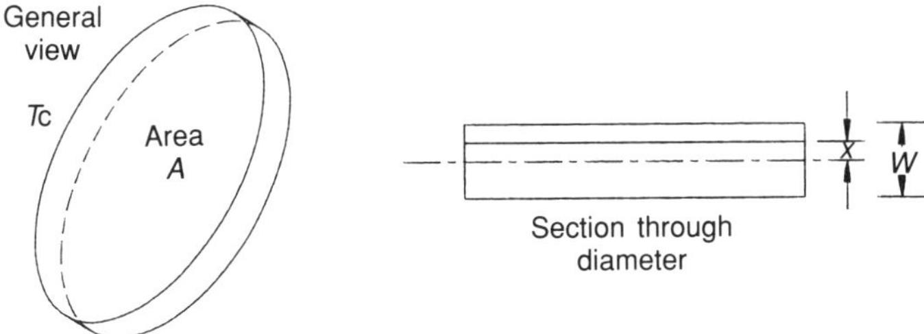

Fig. 1.12 Disc geometry to increase cooled surface area.

equation. High continuous powers can therefore be achieved only with long lasers, which is generally not desirable. Of course if the laser is rapidly pulsed, then equation (1.11) refers to average values, and in that case the power within each pulse can greatly exceed the continuous value for a short time.

1.5.3 Disc geometry

Matters can be improved by building the laser as a series of thin discs of thickness W and area A thus increasing the cooled surface area with respect to laser volume, as shown in Fig. 1.12. The differential equation for this case becomes

$$\mathrm{d}T/\mathrm{d}x = -Qx/K \tag{1.12}$$

where x is the distance from the mid-plane of the disc. The final power equation becomes

$$P = 8\xi A/W(T_{\mathrm{lim}} - T_{\mathrm{c}}) \tag{1.13}$$

where A is the area of one face of the disc, assumed cooled on both faces. In this case the continuous or average power is improved by increasing the area A or reducing the thickness W.

1.5.4 Fluid lasers

To achieve really high powers, it is necessary to convect heat forcibly from the lasing region. This method is applicable only to fluid laser media. Although this analysis is relevant principally to fast-flow CO_2 lasers, it could also be applied to other gas and liquid laser types. Assume a lasing region of length L and cross-sectional area A with the same uniform heat deposition rate Q as used in previous sections. The fluid laser medium flows along the length axis at velocity V propelled by a pumping system external to the laser region. This configuration is shown in Fig. 1.13. It is assumed that conductive and diffusive heat transfer to the boundaries of the lasing

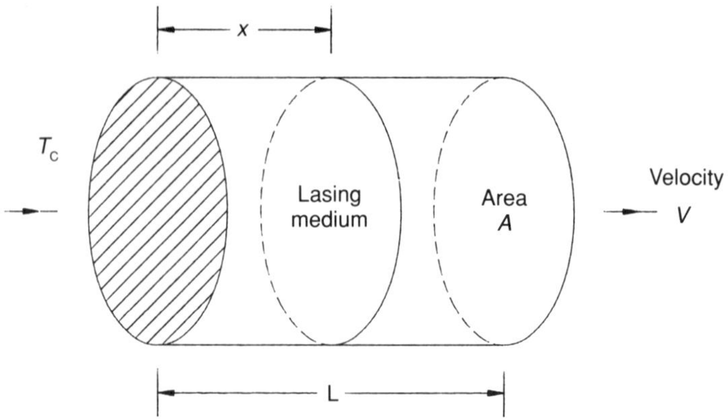

Fig. 1.13 Convection model of laser heat balance.

region are negligible in comparison with the forced convection parallel to the direction of fluid flow.

At a distance x parallel to L along the lasing region, the fluid acquires an amount of heat per unit volume δH where

$$\delta H = Qx/V \tag{1.14}$$

This heat will be converted mainly to a rise in fluid temperature δT such that

$$\delta H = \rho c \delta T \tag{1.15}$$

where ρ and c are the fluid density and specific heat respectively. Equating (1.14) and (1.15) and defining the fluid temperature at the entrance to the lasing region as T_c, we obtain

$$T = Qx/V\rho c + T_c \tag{1.16}$$

The temperature thus increases linearly with distance, reaching a maximum of

$$T_{max} = QL/V\rho c + T_c \tag{1.17}$$

at the exit to the lasing region. Equating this to the limiting temperature and replacing Q by

$$Q = P/\xi AL \tag{1.18}$$

the final power equation becomes

$$P = \xi AV\rho c(T_{lim} - T_c) \tag{1.19}$$

In marked distinction to the cylindrical conduction limited case, the power here increases linearly with both cross-sectional area and fluid velocity.

1.5.5 Worked examples

At the outset it must be stated that the above analyses are only approximate, since they assume constant thermal parameters, constant conversion efficiency and uniform heat deposition. Within these restrictions however, the results are surprisingly useful. The examples cited below come direct from the author's experience with carbon dioxide laser technology. For information on other lasers, the reader is referred to the appropriate chapters in this book.

Conduction limited slow-flow CO_2 laser

The output power for this case is given by equation (1.13), that is

$$P = 4\pi\xi KL(T_{\mathrm{lim}} - T_{\mathrm{c}})$$

setting

$\xi = 0.15$ (15% discharge efficiency)
$K = 0.14$ W/m.K (value for dominant gas He at 0°C)
$T_{\mathrm{lim}} = 250$°C (from statistical mechanics)
$T_{\mathrm{c}} = 0$°C (refrigerated cooling water)

Equation (1.13) yields an output power of 66 W/m. This compares well with experimental data of 50–100 W/m, including the independence on diameter. The calculated figure is based on the limiting temperature being achieved anywhere within a laser cooled only by conduction to the walls. In practice, the laser will continue to operate if parts of it are above this temperature and the laser may produce even higher powers, but the optical quality of the ouput beam could suffer as a consequence.

Forced-convection, fast-flow CO_2 laser

The output power for this case is given by equation (1.19), that is

$$P = \xi AV\rho c(T_{\mathrm{lim}} - T_{\mathrm{c}})$$

It is perhaps more helpful to rewrite this in the form

$$P = \xi\rho c(T_{\mathrm{lim}} - T_{\mathrm{c}})\ \text{Volume/Dwelltime} \tag{1.20}$$

where Volume (= AL) refers to the contents of the lasing region and Dwelltime is the time taken for the lasing gas to be convected across the lasing region. Setting

$\xi = 0.15$ (15% discharge efficiency)
$\rho = 0.05$ kg/m^3 (approximate value)
$c \sim 5$ kJ/kg (approximate value)
$T_{\mathrm{lim}} = 250$°C (from statistical mechanics)

T_c = 0°C (refrigerated coolant in gas heat exchangers)
Volume = 1 m^3
Dwelltime = 2 ms

we obtain an output power of 4.7 $kW/m^3/s$. Putting in typical values for a fast axial flow laser, in which there are four laser tubes each 600 mm long by 25 mm internal diameter, and a dwelltime of about 2 ms, the maximum expected power is 2.8 kW. This is very close to the observed power from real devices.

This form of power equation also gives an interesting insight into methods of scaling to higher powers. Specific dimensions are no longer important; only volume and dwelltime are significant. Thus if technology favours a short length and large cross section (transverse-flow laser), this does not penalize operation. Indeed, in many ways it may be an advantage since low-pressure fans may be used instead of high-pressure blowers as used in axial lasers.

Conclusions

The above examples serve to illustrate how simple theoretical concepts can be used to analyse complicated laser equipment and obtain reasonably useful results. A similar approach is taken to illustrate aspects of processing in the following chapter.

1.6 ACKNOWLEDGMENTS

The author is indebted to Mr P.J. Oakley for helpful and stimulating discussions, to Miss N.A. Hall for preparation of the manuscript, to Mr P.D. Evans for assistance with the illustrations, and to TWI for permission to publish this material.

REFERENCES

Einstein, A. (1917) *Phys. Zeit.*, **18**, 121.
Fox, A.G. and Li, T. (1961) *Bell System Technical Journal*, **40**, 453.
Geusic, J.E., Marcos, H.W. and van Uitert, L.G. (1964) *App. Phys. Letts*, **4**, 182.
Gordon, J.P., Zeiger, H.J. and Townes, C.H. (1955) *Phys. Rev.*, **95**, 282.
Javan, A., Bennett, W.R. and Herriott, D.R. (1961) *Phys. Rev. Letts*, **6**, 106.
Maiman, T.H. (1960) *Nature*, **187**, 493.
Patel, C.K.N. (1964) *Phys. Rev.*, **136A**, 1187.

2

Background to laser processing

R.C. Crafer and P.J. Oakley

2.1 INTRODUCTION

Laser processing comprises five major activities, namely:

1. Generation of light by a laser;
2. Transmission to the workpiece;
3. Beam monitoring and quality assurance;
4. Interaction with the workpiece;
5. The process.

Although each activity can be and often is considered in isolation, one should ideally consider all five as an overall process chain. With this in mind, this chapter examines each activity from a process engineering viewpoint and takes its examples from continuous power welding and surfacing using continuous carbon dioxide lasers.

2.2 GENERATION

The high intensity light beams used for materials processing are generated by lasers as described in Chapter 1. From a process viewpoint the laser can be regarded as an energy transformer, taking in high-power, low-grade energy from a **pump** and emitting lower-power, higher-grade energy as a laser beam. As with all heat transformers, some energy is lost in transformation, and manifests itself as heat in the laser device. It is this heat which effectively limits the performance of industrial lasers.

There are two major industrial laser types. The Neodymium YAG laser (YAG for short) takes as its source of low-grade energy the light from a high-intensity flashlamp. This is transformed by a rod of Yttrium Aluminium Garnet doped with triply ionized Neodymium ions into a population inversion at a wavelength of 1.06 μm. In the case of the Carbon Dioxide laser, the low-grade energy is provided by an electrical supply. The

Laser Processing in Manufacturing. Edited by R.C. Crafer and P.J. Oakley.
Published in 1993 by Chapman & Hall, London. ISBN 0 412 41520 8

transformation occurs in a low-pressure mixture of carbon dioxide, nitrogen and helium gases, producing a population inversion at a wavelength of 10.6 μm. In both cases a pair of mirrors fashions the stimulated output into a collimated monochromatic light beam. Figure 4.12 illustrates a commercial YAG laser, and Fig. 7.1 illustrates a commercial carbon dioxide laser.

2.3 BEAM TRANSMISSION SYSTEMS

The purpose of the beam transmission system is to relay the high-power beam from the laser generator to the workpiece in a manner appropriate to the application. The transmission is probably the most critical aspect of the laser processing chain, and applications can succeed or fail depending on the suitability and efficiency of the transmission. A proper study of transmission systems begins with the laser generator itself. There are several methods by which the beam power can be coupled out of the laser generator and launched towards the workpiece. All known methods somewhat distort and perturb the beam thereby reducing its potential effectiveness for the intended application.

At powers up to 1 kW in YAG and 5 kW in CO_2, lasers are normally fitted with a **stable** resonator, from which the beam is extracted through a partially reflecting transparent plane window. With YAG lasers, the window material is glass, which can be produced with very low absorptions at the 1.06 μm wavelength. For reasons related to the long infrared wavelengths involved, the window material for carbon dioxide lasers is a yellow transparent material called zinc selenide. Being a semiconductor it exhibits strong absorption of light when heated, resulting in thermal lensing or, in extreme cases, mechanical failure. The actual temperature obtained is a balance between heat absorption from the beam and heat losses to the surroundings. Absorption occurs via impurities within both substrate material and reflective or antireflective coatings, and also by electronic absorption involving conduction band electrons. Cooling occurs predominantly via convection at the window faces and conduction into a cooling medium at the edges. In practice, by judicious choice of material and coatings for the size and power of laser, beam distortion can be minimized, certainly at the powers quoted above, but long-term effects such as surface contamination can cause unexpected absorption, and such possibilities should always be anticipated in maintenance schedules.

At powers above 5 kW in CO_2, it is common practice to use an **unstable** resonator. Here, coupling is achieved via mirrors, which being usually metallic and of high thermal conductivity can handle much higher powers than semiconductor windows such as zinc selenide. However it is still necessary to transmit the beam from the generator environment to the working environment, and this requires some form of window. The preferred type of window in this case is the aerodynamic window, where the beam is brought to a preliminary focus through a series of differentially pumped apertures and subsequently reformed, using off-axis optics.

Having extracted the beam from the laser, it must be directed to the workpiece. For YAG laser applications, the currently preferred method is the flexible optical fibre; this provides remarkable flexibility using relatively simple robotics, and is having a dramatic impact on the potential uses of this type of laser. Further details can be found in Chapter 4. For CO_2 lasers, where fibre technology is not adequately developed, there are four basic variations of a mirror theme:

1. Moving workpiece. Suitable for small, low inertia workpieces or for large and immobile lasers. Advantages are simplicity of construction and constancy of beam path, allowing the designer to minimize transmission distortion.
2. Moving laser – moving mirrors. Laser and mirrors move in unison. Suitable for low-power, low-inertia lasers with large workpieces. Advantages are simplicity of construction and constancy of beam path, allowing the designer to minimize distortion. Optically similar to the moving workpiece.
3. Moving mirrors. This is almost the ideal solution since it places no limitations on the laser generator or the workpiece, but poses difficult problems for both the mechanical and optical designers. Many designs have been attempted, all with their own advantages and problems. Perhaps the most exciting recent development is the robot-driven articulated beam tube. This is the nearest approach so far to a flexible cable capable of carrying a multi-kilowatt beam from generator to workpiece.
4. Moving mirrors – moving workpiece. A solution which often achieves the objectives of the moving mirror approach with reduced complication.

Once in the vicinity of the workpiece, the beam has to be fashioned into a form suitable for the application. For welding and cutting at low powers, lenses are commonly used to bring the beam to a focus. For high-power welding and cutting applications, mirrors are usually employed as focusing elements. For technical reasons, off-axis operation is commonly employed which sometimes requires the use of specialized aspherical shapes, such as the off-axis paraboloid.

For heat treatment applications, there are three types of solution, namely defocusing, beam redistribution and raster scanning. Beam redistribution involves juxtaposing and modifying different parts of the unfocused beam to form an intensity profile which provides an approximately uniform treatment depth when the profile is moved across the workpiece. Special multifacet mirrors or aspheric components such as axicons are used to accomplish this. Raster scanning involves moving (vibrating or rotating) one or more of the transmission mirrors so that a conventionally focused spot traces out a raster pattern of the required time averaged intensity.

Many useful transmission systems have been constructed but there are also many problems to be avoided. Every optical element in the beam transmission system reduces the quality and quantity of the beam reaching the workpiece. The loss per mirror may be only a few per cent, but in a multimirror system, such losses will quickly accumulate. With the exception of perfectly figured aspherical mirrors, all other curved mirrors produce aberrations. Lenses, in addition to these problems, also reflect an amount of light from each surface, although this may be reduced to acceptable levels on some materials by anti-reflection coatings. In addition, if any of the components becomes heated by the beam, then its shape will alter and the beam quality will become further reduced. Finally, in some high power applications, the atmosphere between the optical elements may itself become heated and reduce further the quality of the beam. This effect is known as thermal blooming.

For a more detailed discussion of the optical considerations of beam transmission systems, the reader is referred to Chapter 3.

2.4 BEAM MONITORING

2.4.1 General

The beam emitted by a laser can be characterized by three parameters: power, diameter and intensity distribution. Beam power is measured routinely in lasers used for materials processing. It is a principal variable for processes such as welding, cutting, surfacing, etc. Beam diameter is defined by the optical resonator; however, measurement of a partially or fully focused beam is also of interest. The intensity distribution within the beam is a function both of the resonator and the optical components in the beam train. In the transmission the distribution is usually axisymmetric and can be measured by the beam profile along a radius or diameter. Given the beam profile, a diameter may be deduced either in terms of the fraction of beam power passing within it, or in terms of points where the intensity is reduced to a fraction of its peak value. Although making fine adjustments to the laser optics by reference to intensity distribution as well as laser power is desirable, this is not achieved easily with present measurement techniques. As lasers become established as production machines for a range of applications, the monitoring of these parameters will become increasingly important for process and quality control.

Beam parameters can be measured using the total beam, or by sampling a known fraction of the beam. Total beam measurement is the simplest technique but has the disadvantage that measurements are not available during processing. Three methods of sampling part of the beam, usually a

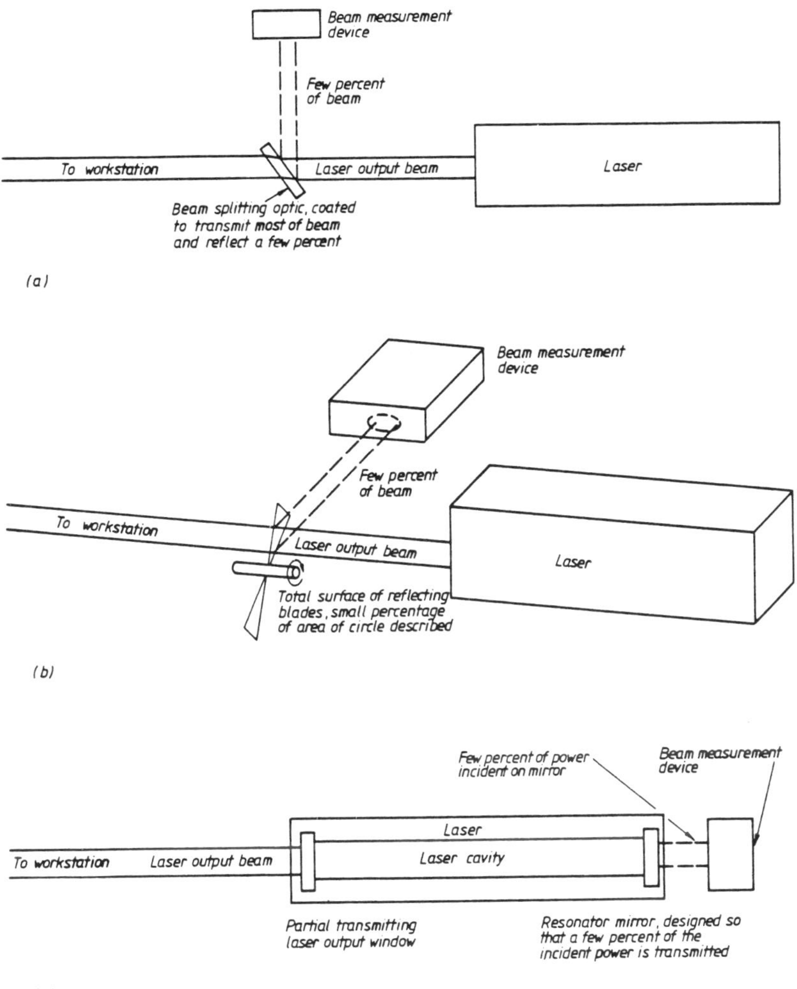

Fig. 2.1 Methods of sampling a laser beam. (a) static beam splitter, (b) rotating chopper mirror, (c) leaky resonator mirror.

few per cent, are shown in Fig. 2.1. As only a small sample of the beam is monitored, techniques with lower power capabilities can be used. All three allow in-process monitoring, but reduce the accuracy of measurement because of uncertainties in the fraction sampled. Each method has additional disadvantages. A static beam splitter (Fig. 2.1a) is a delicate component and can be expensive. It will absorb some power from the transmitted beam, and may not be suitable for use at higher power densities. The chopper wheel (Fig. 2.1b), which consists typically of a rotating

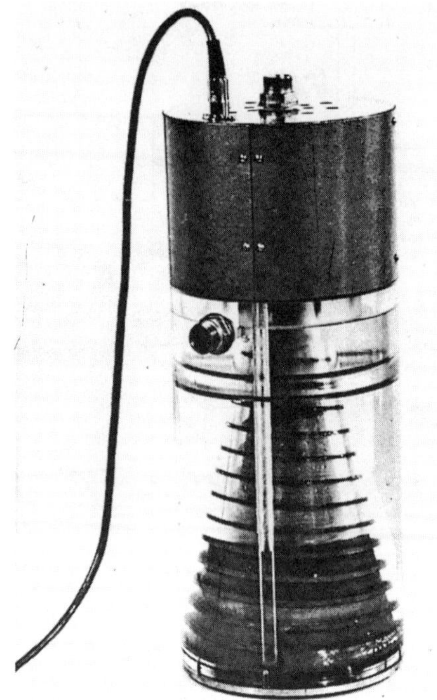

Fig. 2.2 Water-cooled calorimeter.

45° gold-plated copper blade, cuts across the beam and hence there is a modulation of the laser beam which may be significant when processing material at high speeds. The use of a coated dielectric mirror, or a small aperture in a conventional mirror at the non-output end of a laser resonator, will also sample a fraction of the beam (Fig. 2.1c). A coated dielectric mirror is again an expensive and delicate component and a restriction on maximum allowable power density may be encountered. An aperture in a conventional fully reflecting mirror necessitates a known intensity profile within the resonator if an accurate fraction is to be sampled. This third technique cannot of course be applied to lasers operating in an oscillator–amplifier configuration.

Total beam power is measured usually by calorimeters. Air-cooled types are restricted to lower powers than those cooled by water. Figure 2.2 shows a water-cooled calorimeter in which the beam is absorbed on an internal conical surface. Low powers up to 1 kW can be measured on water-cooled calorimeters with a flat absorbing surface. Laser systems are usually purchased with a calorimeter; however, free-standing calorimeters are also commercially available. Measurements can be made either by directing the total beam into a calorimeter, or by sampling a known fraction of the beam. When measuring the total beam, a mirror is used to switch

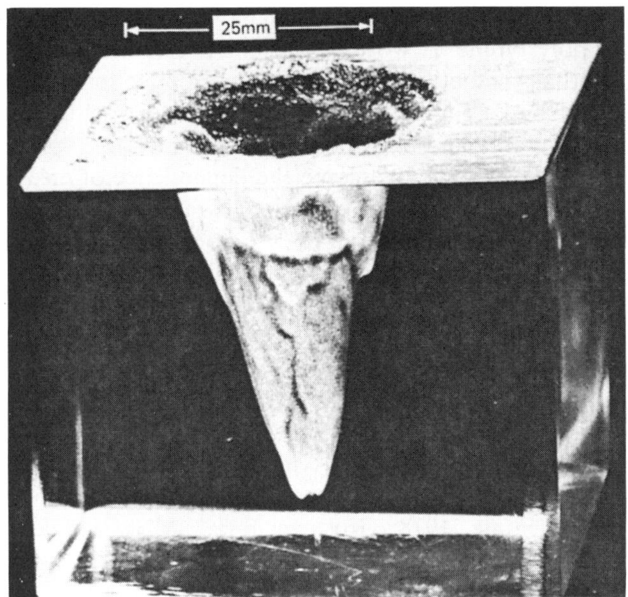

Fig. 2.3 Deep burn print in acrylic block.

the beam from the workstation to the calorimeter. The calorimeter therefore provides a beam dump while the laser is operating but not processing. A combination of in-process measurement and total-beam calorimetry provides a comprehensive power measuring system for laser processing.

2.4.2 Diameter and intensity distribution

Heating or charring sensitive materials is the simplest method of looking at the profile of an unfocused or partially focused beam. Materials used for this include wood and firebrick. This technique gives an indication of beam diameter and symmetry. It is, however, a technique where the affected area, and hence the measured beam diameter depends on both power and exposure time. It is therefore more suitable for comparative measurements. Courtney and Steen (1978) describe this technique in detail. Incandescant mode plates and materials that are chemically sensitive to the laser wavelength or its heating effects can also be used with low-power lasers or small fractions of high-power beams.

Measurements made by vaporizing acrylic or similar materials are also subject to reservation; however, more information can be obtained than with heating or charring because of the possibility of vaporizing a deep region in the acrylic and hence making a three-dimensional print. A deep acrylic print of an unfocused beam is shown in Fig. 2.3. This type of print gives a better indication of beam diameter and intensity distribution than a shallow print. Meyerhofer (1968) has analyzed the case of deep prints in expanded polystyrene. A word of warning is essential at this point. All plastics and other synthetic materials emit poisonous fumes when heated by a laser beam, and great care should be taken to extract the fumes safely from the working area and to dispose of them in an appropriate manner.

Courtney and Steen have overcome the dependence of measured beam diameter on laser power and exposure time by measuring the time for the temperature of a laser heat spot to rise to a fixed percentage of its equilibrium value. This time is compared with predictions for a guassian beam, and an effective gaussian beam diameter deduced. Pyroelectric detectors can be used to obtain intensity profiles. As they respond only to variations in power and not to power itself, it is necessary to move the device through the beam or to use a chopper wheel. The devices are fragile, expensive and susceptible to thermal overload, and are therefore limited to lower power densities – typically 1 W/mm^2. Various sophisticated systems are available combining pyroelectric detectors with beam scanning optics (Foulk, 1978; Grosjean *et al.*, 1978). Sepold (1980) has described the use of a Nipkow disc technique with a 5 kW CO_2 laser. Pyroelectric vidicons have been used for quasi real-time measurement of intensity profiles. These devices are television tubes with pyroelectric cathodes.

Beam diameters can also be measured by their effect on a workpiece, and this could indeed be regarded as the ultimate monitor. Partially focused beams are used to effect solid phase transformations or melting in laser surfacing processes. Where melting has occurred, the treated width can be measured directly on the surface. A more accurate method however is to take a measurement of an etched and polished cross section of a heated specimen (Chapter 10. Shallow tracks made by traversing an angled acrylic sheet across the laser beam will give the position of focus, but because of the high power density, no information about the beam diameter can be deduced close to focus.

Of particular interest in welding and cutting applications are the focal diameter and profile of the beam. Two methods are described by Crafer and Oakley (1981) and Lim and Steen (1982). The devices involved are complex and the results require interpretation. Interested readers should consult the references. A typical signal from one such scanner is shown in Fig. 2.4. This signal is derived from a three-dimensional beam slice (Fig. 2.5) and is used to reconstruct the two-dimensional intensity profile using a mathematical process known as deconvolution. A similar problem arises when monitoring electron beams, and a technique based on a method due to Harker (1975) has been employed in this case. This method is tolerant to noise and signal inaccuracies but requires a narrow sample width, typically 20:1 beam diameter:sample width, to permit accurate beam intensity profile reconstruction. In the laser case it is more difficult to use such a

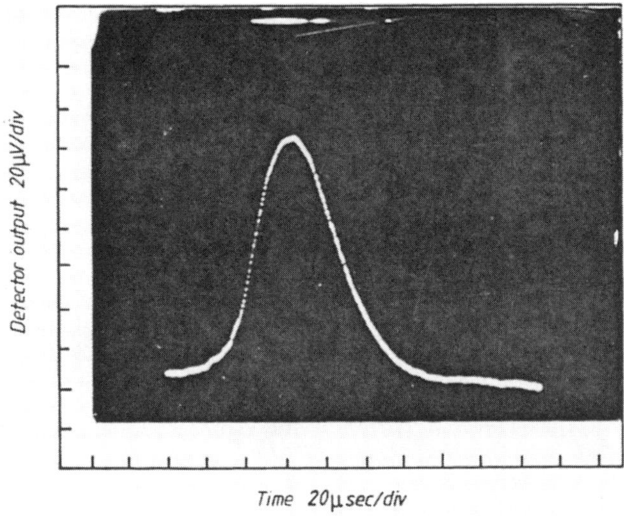

Fig. 2.4 Oscillogram from high-power laser beam scanner.

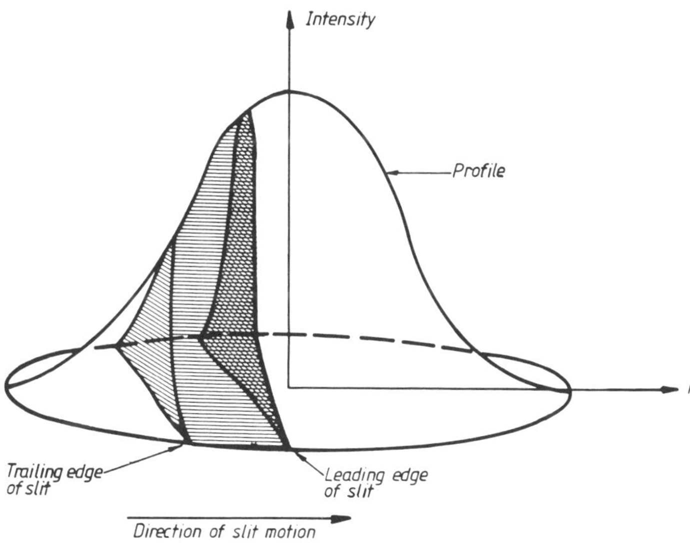

Fig. 2.5 Geometry of beam scanner interaction. Detector signal is proportional to shaded volume. The 3D shape is the intensity distribution. The 2D curve through the axis is the intensity profile.

high ratio because the low signal levels become obscured by noise, thus ruling out the method. An alternative, more limited method of interpretation is sometimes used in which 'reference' results are compared with actual displays, and the most appropriate reference chosen. Recently, improved devices have appeared on the market for monitoring high-power beams which overcome many of the limitations mentioned above.

2.5 LASER–WORKPIECE INTERACTION

First of all, the beam strikes the material surface. In the case of a non-metal, the crystal lattice composed of positive and negative ions becomes polarized by the electric field of the beam. Upon reversal of the field, the pent-up energy is released and helps the beam to propagate further. Depending on the electrical 'stiffness' of the material, some energy propagates forwards, and some is reflected backwards. The release of energy into the transmitted or reflected beams occurs without loss, and accounts for the excellent behaviour of window materials such as quartz, glass and zinc selenide. In the case of metals, which are crystal lattices filled with electrons, the electrons gain energy from the electric field, and transfer it to lattice impurities thereby heating up the metal. Analysis shows that most of this absorption occurs in a narrow surface layer a few atoms thick, so that laser absorption in metals is truly a surface effect. The absorbed fraction may vary from about 1% in copper to tens of percent for steels. A variation of this absorption process is used for solid-phase surface treatment or surface alloying. Clearly even a few tens of percent absorption is insufficient, so a matt surface coating such as colloidal graphite is applied. Light not absorbed by any one particle is likely to be reflected on to other particles and so on, thus greatly enhancing the absorption process. The heat absorbed by the coating is then transferred to the metal surface by conduction. When sufficient heat has been absorbed, the coating burns away completely and terminates the process.

For welding, a second stage of absorption is required as coatings cannot be used, because they would either be incorporated into the weld metal with possible adverse characteristics or be burned off in an uncontrollable fashion. In this case a prime requirement is that the first stage of absorption is able to bring the metal surface to its boiling point. Low-absorption metals such as gold and copper will clearly be more difficult to vaporize than high-absorption steels, and indeed at present only YAG lasers with their relatively short wavelengths are capable of welding these materials. Once vaporized, some of the metal electrons become free, a process called ionization. These free electrons absorb energy directly from the incoming light beam by a process known as Inverse Bremsstrahlung, resulting in higher temperatures, more ionization and increasing absorption. The rapidly increased absorption vaporizes the surface, which recedes and produces

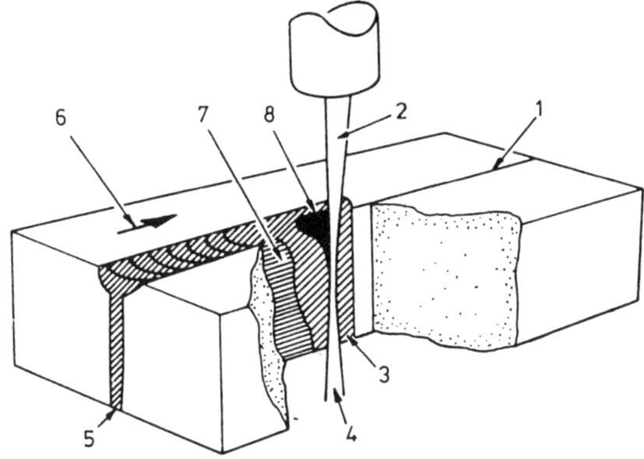

Fig. 2.6 Deep penetration keyhole. (1) close butt joint, (2) laser beam, (3) molten metal, (4) proportion of power passes through keyhole, (5) full penetration weld, (6) welding direction, (7) solid weld bead, (8) keyhole.

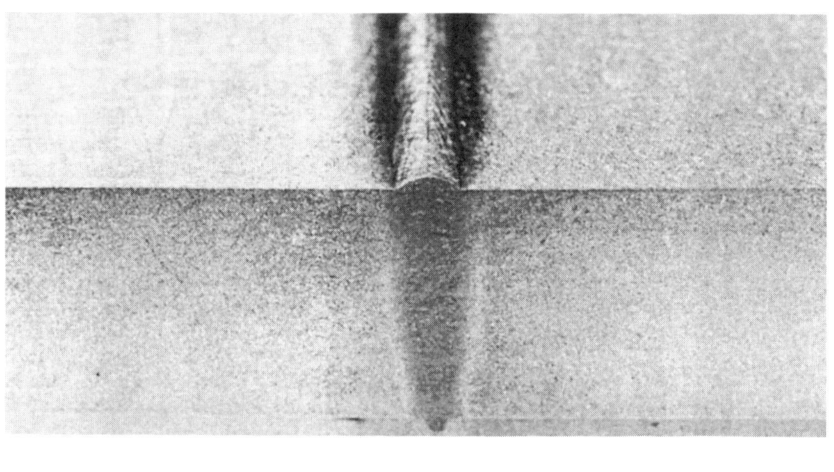

← 10 mm →

Fig. 2.7 Metallographic section of laser weld.

a cavity through the depth of the metal. This cavity is known as a keyhole and forms the basis for deep penetration keyhole welding (Figs 2.6 and 2.7). The keyhole is a liquid-lined cavity with a very hot ionized gas core. Surface tension and gas pressure hold the liquid in place against gravity and other forces, while light absorption in the central gas core keeps the process going.

Welding takes place by moving the beam, and hence the keyhole, through the metal and melting the adjacent material. In simple terms, the fluid flow of molten metal past the keyhole can be ascribed to an imbalance of surface tension forces, but for a more detailed analysis the reader is referred to Dowden *et al.* (1985a, 1985b). In practice, keyhole absorption is very efficient and accounts for the majority of the beam power. In extreme cases, the hot metal vapours escaping from the keyhole can actually absorb the beam above the metal surface, causing keyhole collapse and poor welding performance. This is known as the plasma effect, and various methods have been used to overcome it, principally the use of auxiliary gas jets to suppress and remove the absorbing plasma cloud.

Lasers between 100 W and 10 kW are becoming increasingly accepted as production tools for welding. The highest-power machines are capable of welding, in one pass, thicknesses of at least 25 mm of steel while, at the other end of the scale, metals of foil thicknesses can be welded using lower powers. This makes the process suitable for a wide range of applications, particularly where numbers of similar assemblies are to be manufactured, allowing full automation of the process and very high production rates. Actual and potential applications range from electronics to heavy fabrication, although at present the majority of applications require small weld penetration depths. Applications are also restricted by the materials, with the higher-reflectivity metals such as aluminium alloys, and copper, being difficult or impossible to weld with a CO_2 laser, but quite feasible with a YAG laser.

2.6 PROCESS PARAMETERS

Six process variables can be identified. These are power, pulse waveform, traverse speed, diameter of the focused beam, depth of focus, and type of gas shielding. The two beam parameters are interrelated, both being dependent on the focusing system used. The type of laser used dictates other variables such as the wavelength of the beam and intensity distribution. Sophistications of the process, such as the use of wire feed or beam oscillation, introduce further variables which require quantifying. Measurement of the spatial beam parameters (power, diameter, intensity distribution) have been described earlier in this chapter. For a laser to be able to process efficiently, it is necessary to operate with the beam focused at or close to the workpiece surface. The diameter of the focused beam and the depth of focus quantify the beam in the focused region and, whether based on lenses or mirrors, these parameters are related to the f number of the focusing system. For any particular beam diameter, this means that these parameters are related to the focal length of the focusing system. Although researchers have investigated the effects of different focal lengths on process performance and quality for particular lasers, there is insufficient data available to formulate generalized rules (Dawes, 1983).

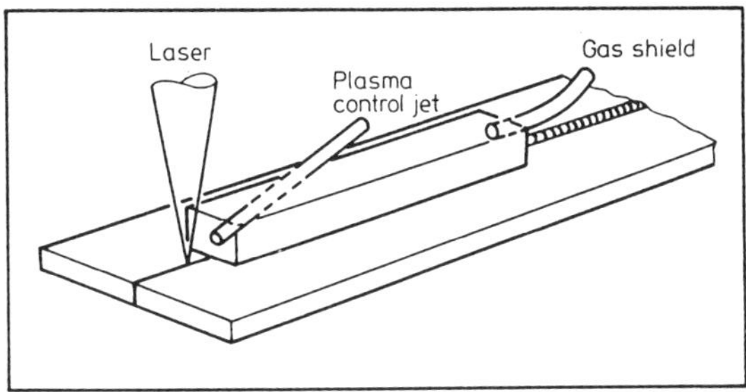

Fig. 2.8 Schematic of plasma control device.

It is, however, possible to make some comments. A focusing system with a short focal length will result in a smaller diameter of focused beam than a system with a long focal length. However, the penalty for using a short focal length is a much reduced depth of focus. Short focal lengths (<150 mm with a 30 mm diameter beam) are therefore more suited to the welding of thin materials (<4 mm) where high-power density is required to maintain the weld keyhole at high speed. The greater depth of focus of a longer focal length system is more efficient for welding thicker materials at comparatively slow speeds. Practical considerations of access with very short or very long focal length systems, and the vulnerability of short focal length systems to damage by reflections or spatter from the workpiece have also to be taken into account when selecting the focusing arrangement. In the latter respect, mirrors are usually better than lenses.

The gas shield has several functions. It protects optical components from fume and spatter, ensures effective transmission of the beam through the hot region of gas just above the workpiece, and protects the molten material from reaction with the atmosphere. It is generally agreed that helium is an effective gas to use, but it is expensive, so there is much work currently in hand to investigate other gas mixtures (Jimbou *et al.*, 1980; Shinada *et al.*, 1981). The level of gas shielding required is dependent on the specific application, but in general the slower the speed, the more opportunity there is for the atmosphere to react with the weld metal, so a more comprehensive gas shield is required. The gas shield must also provide plasma control, particularly at high powers and low welding speeds. At low welding speeds a plasma of metal vapour escaping form the top of the keyhole interacts with the laser beam to lessen the penetration capability of the beam. This is a severe problem only with CO_2 lasers. At the Nd:YAG wavelength, the effect is 100 times less serious. A high-velocity jet of helium is usually directed at the top of the weld to blow away this plasma (Pauley and Russel, 1978). Figure 2.8 shows in schematic form a plasma control device. Another technique that has been proposed (Banas, 1989) is to pulse the laser beam at a high frequency so that the duration of each pulse is shorter than the time taken to generate the plasma, thus enabling the beam to penetrate the material efficiently.

2.7 SURFACING

Lasers are inefficient bulk heating devices and their use for through thickness heat treatment has not proved feasible. However, they are very effective at heating discrete areas very rapidly, and are therefore suitable for certain heat treatment and surfacing application. The laser can produce a high power density, matched only by electron beam equipment. When using a conventional heat source, such as an oxy-acetylene flame for surface treatments, the relatively low power density allows heat to be conducted into the bulk of the component while the surface is reaching the required temperature, thus heating the surface inefficiently. The high power density laser beam heats the surface much more rapidly, reducing time for heat conduction into the bulk of the component. The laser is localized controllable heat source, and can therefore be utilized to treat a component efficiently by heating only discrete areas. The processes with which laser heat treatment and sufacing techniques must compete include a wide range of comparatively low-cost conventional processes. The use of laser techniques must therefore offer significant advantages. The generally agreed advantages are:

1. Treatment can be localized to the required area;
2. Heat input is relatively low, giving minimum thermal distortion of the component;
3. Little or no post-treatment machining is required;
4. Complex component shapes can be treated;
5. The laser beam can be directed, by mirrors or fibres, to treat inaccessible areas of components;
6. Most treatments are rapid and can be incorporated into a production line.

Specific advantages can be found with each of the different laser surfacing processes, and secondary, sometimes unexpected, benefits have been found when these processes are considered for a particular application.

2.8 PROCESS CLASSIFICATION

Laser surface treatments have been classified into three types, namely those involving heating, melting or shocking, Fig. 2.9 (Gnanamuthu, 1979). The processes can also be divided into those relying on a metallurgical change in the surface of the bulk material – transformation hardening, annealing, grain refining, glazing and shock hardening, and those involving a modification of the chemical composition of the surface by addition of

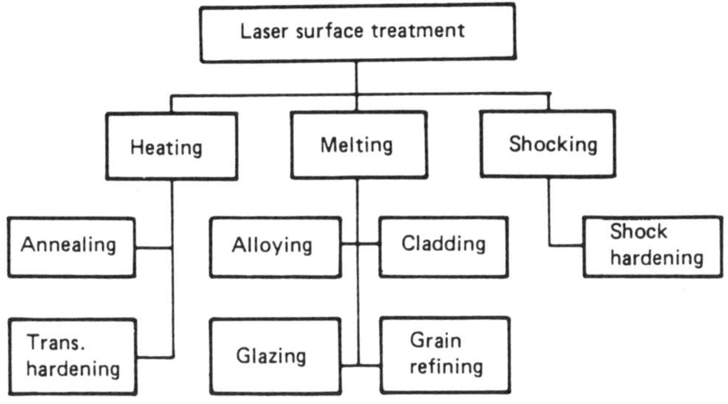

Fig. 2.9 Methods of laser surface treatment (after reference (16)).

new material – alloying and cladding. Subsequent to these classifications an additional technique called particle injection has become the subject of research. This involves the addition of solid powder into the molten surface of the substrate, and hence can be classified with alloying and cladding.

Unlike laser welding and cutting, where the beam diameter is focused to a minimum size at the workpiece surface in order to give a narrow fusion and heat affected zone, most laser surfacing processes require a reduction in power density and preferably treat a wider band of the component. A range of beam handling techniques from defocusing to rastering and multifaceted mirrors can be used. The use of a defocused beam is the simplest way to reduce power density and give the required treatment width. However, it is difficult to obtain a uniform power density across the beam. A defocused gaussian beam, for example, will have maximum power density at its centre and thus will produce a maximum treatment depth at the centre of its track across a component. A more complex beam handling system can be utilized to give a uniform power density, but this usually is reflected in higher cost optical components, together with less flexibility and reliability.

Metal surfaces are good reflectors, particularly at the 10.6 μm wavelength of CO_2 lasers. This high reflectivity, and hence poor coupling of the energy to the workpiece, can limit the effectiveness of laser metalworking processes. The absorption of infrared radiation does, however, increase with increasing surface temperature and shows a sharp increase once the material is molten. This reflectivity is overcome when laser welding and cutting by the beam being absorbed in a plasma-filled keyhole. During surfacing processes, where the component surface is as undisturbed as possible and its temperature is lower, it may be necessary to treat the component surface to increase its absorption of the laser beam, particularly for processes that do not involve surface melting. To be effective, the coating must be very thin to allow effective transfer of energy to the component. Typically used coating materials are matt black paint, colloidal graphite, and zinc or manganese phosphates. Powdered materials on a component surface will facilitate the absorption of the laser beam and therefore if material is added as a powder in alloying or cladding, the application of a coating to enhance absorption may not be required. Many of these processes will be considered in later chapters, and interested readers are referred to references (17) to (30).

2.9 HEAT-FLOW MODELS

Heat flow within the process zone is difficult to model accurately. Most successful models to date have utilized techniques such as finite element or finite difference analyses with parameters varying both spatially and temporally. Although such models are capable of considerable accuracy, they are cumbersome, and require extensive computing facilities. If approximate results only are required, then certain simple analyses may be employed, provided the user is aware of their limitations. Four steady-state models will be considered here as follows:

1. Surface heating of thick solid by stationary point source (conduction limited spot welding);
2. Surface heating of thick solid by moving point source (conduction limited seam welding);
3. Surface heating of thick solid by stationary disc source (surface treatment and spot welding);
4. Through depth heating of solid by moving line or moving cylinder source (deep penetration welding).

2.9.1 Surface heating of thick solid by stationary point source

Consider a point source of strength Q_{point} (power) buried within an infinite solid. In the steady state, heat crossing any spherical boundary centred on the source must equal Q_{point}. The surface area of a spherical boundary of radius r is $4\pi r^2$, so the heat crossing it is, by Fourier's Law:

$$\text{Heat} = -K.4\pi r_2.\text{d}T/\text{d}r \tag{2.1}$$

where K and T are the thermal conductivity and the temperature respectively. Equating (2.1) to Q_{point} we obtain

$$Q_{point} = 4\pi Kr.\text{d}T/\text{d}r \tag{2.2}$$

Separating variables and solving for T

$$T = Q_{point}/4\pi r + T_{amb} \tag{2.3}$$

where T_{amb} is the ambient temperature.

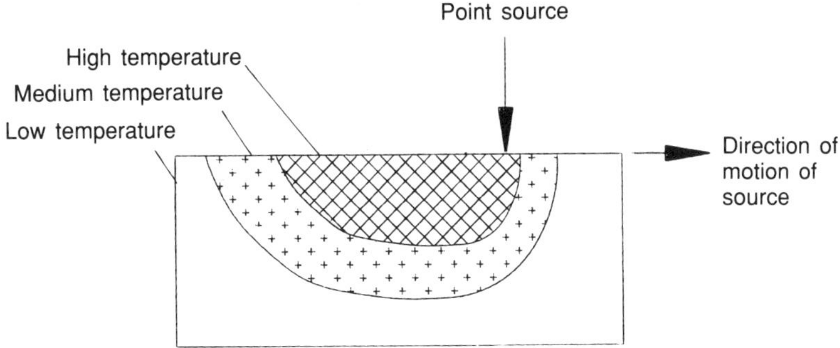

Fig. 2.10 Isotherms for moving point source.

For surface heating, the heat Q_{point} from the source is dissipated over only half of the material volume, so the term containing the heat source in (2.3) must be doubled, hence

$$T = Q_{point}/2\pi r + T_{amb} \tag{2.4}$$

This equation is also relevant to molten material provided that the conductivity K is sensibly constant, and that the time scale of the interaction is sufficiently short to rule out convective heat transfer.

2.9.2 Surface heating of thick solid by moving point source

There is no simple analysis for this case, so we must instead resort to an extension of the full theory which is described in standard texts on heat flow, such as the well known work by Carslaw and Jaeger (1959). The steady state solution is

$$T = Q_{point}/2\pi Kr.\exp[-U(r - x)/2k] + T_{amb} \tag{2.5}$$

where U is the velocity of the heat source, x is a distance in the direction of travel, r is a distance in any direction from the instantaneous heat source, and k is the thermal diffusivity. Notice how this collapses to the stationary case value given by (2.4) when the velocity U becomes zero. The approximate form of the isotherms is given in Fig. 2.10. Since these are similar along planes parallel to the source velocity, two things in particular are noticeable:

1. The widest part of the thermal profile is behind the source.
2. The thermal gradient is increased in front of the source and diminished behind it.

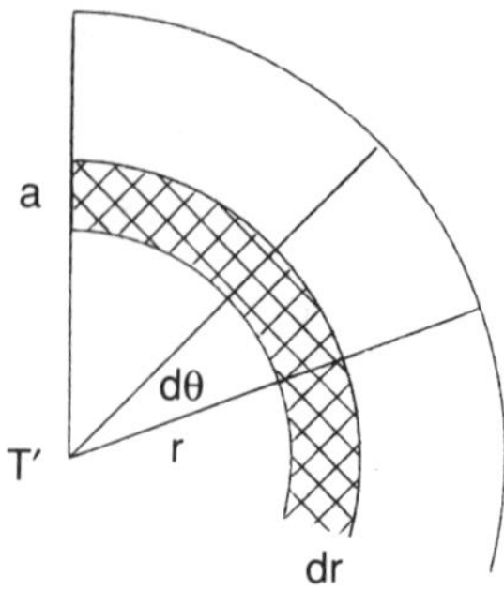

Fig. 2.11 Disc source geometry for calculating temperature distribution.

2.9.3 Surface heating of thick solid by stationary disc source

As previously, we perform the calculation for the full infinite solid and then double the source dependent part of the result. Let the absorbed power density at the surface be a constant uniform value Q_{uniform} (power per unit area). The temperature rise dT' at the centre of the surface of the heated disc due to heat conducted from an elementary annulus of radius r as shown in Fig. 2.11 is

$$dT' = Q_{\text{uniform}}/4\pi K.d\theta.dr \tag{2.6}$$

The full temperature rise T' is then given by integrating the expression in (2.6) over θ and r, which gives

$$T' = Q_{\text{uniform}}a/2K \tag{2.7}$$

Setting the total absorbed power P to $\pi a^2 Q_{\text{uniform}}$ for a uniform disc source, and $T' = T - T_{\text{amb}}$, and doubling the source-dependent term for the infinite solid gives

$$T_{\text{uniform}} = P/\pi aK + T_{\text{amb}} \tag{2.8}$$

Equation (2.8) enables an estimate of surface temperature to be made for a uniformly heated laser spot. A similar calculation for a gaussian beam profile of spot size equal to the uniform disc above yields a marginally different expression

$$T_{\text{gaussian}} = 2P/\pi^2 aK + T_{\text{amb}} \tag{2.9}$$

2.9.4 Through depth heating of solid by moving line or moving cylinder source

This case, depicted in Fig. 2.12, differs from the previous three in that the dominant heat loss mechanism is assumed not to be conduction,

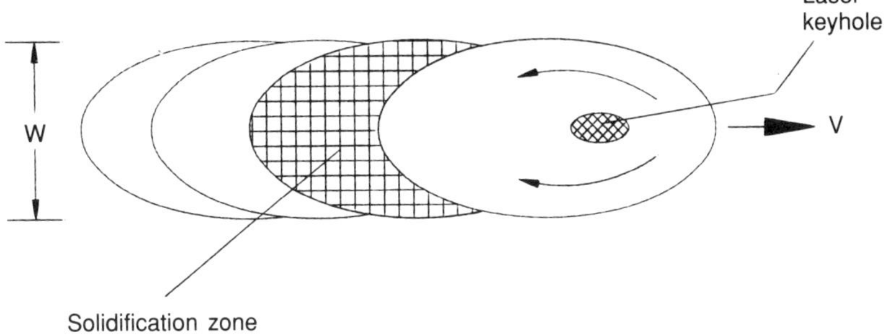

Fig. 2.12 Plan view of laser weld.

but convection by hot liquid metal out of the heated zone. This rather sweeping assumption is justified by the observations that:

1. In practice, melting is a prerequisite of this type of heating.
2. Analytically, a steady-state solution cannot be achieved by solid or liquid conduction alone.

The line or cylinder source of strength Q_{line} (power per unit length) transfers heat to the molten zone, and by its effects, ensures a flow of molten metal from right to left in the diagram. This liquid flow removes heat from the interaction zone at rate $v\rho cT_{\text{av}}W\text{d}z$ where v is the line source velocity, c the specific heat of the material, T_{av} the average temperature, W the width of the molten zone and dz is a length along the line source.

Equating this to the heat input rate $Q_{\text{line}}\text{d}z$ we obtain

$$W = Q_{\text{line}}/v\rho cT_{\text{av}} \tag{2.10}$$

which is the controlling equation for a laser keyhole.

2.9.5 Discussion

All four analytical models serve to illustrate particular aspects of laser processing. The stationary point source model, equation (2.4) illustrates the thermal distribution within a laser-heated spot at distances from the surface exceeding the size of the focused laser spot. This restriction is purely mathematical and has no technological significance. A more relevant case is the moving point source, equation (2.5). This shows the effect of relative motion of source and workpiece, and demonstrates the distortion of the isotherms. At higher powers the surface vaporizes leading to keyhole formation. The critical power and focused spot size required for this can be estimated from equations (2.8) or (2.9) depending on whether the focused distribution is nearer to uniform or gaussian. Having generated a keyhole, the temperature distribution changes from hemisperical to cylindrical. The reduced surface area and shorter thermal cycle time mean that conduction ceases to be the dominant mechanism and is supplanted by convection, giving rise to the keyhole equation (2.10).

2.9.6 Worked examples

1. To estimate the laser power required to vaporize the metal surface for keyhole initiation. In this case we assume that the laser interaction with the surface can be modelled in a gaussian manner, as described by equation (2.9)

$$T_{\text{gaussian}} = 2P/\pi^2 aK + T_{\text{amb}} \tag{2.9}$$

   Assume the following parameters:

   Focused spot radius (a) = 250 μm (typical of industrial CO_2 lasers)
   T_{gaussian} = 2750°C (boiling point for mild steel)
   Thermal conductivity (K) = 30 W/m/K (= value just below melting point)
   Absorption of laser beam = 10% (estimated)

   Applying these values gives a predicted power around 700 W, which is not untypical for the welding of mild steel with a CO_2 laser.
2. To estimate the width of molten (or weld) zone for a 3 kW, CO_2 laser weld in 5 mm thick mild steel using the keyhole equation (2.10) and assuming the following parameter values:

   $Q_{\text{line}} = 6 \times 10^5$ W/m (3 kW absorbed over 5 mm depth)
   $V = 1.5 \times 10^{-2}$ m/s (typical achieved value)
   $\rho = 8 \times 10^3$ kg/m^3 (typical density for steels)
   $c = 6 \times 10^2$ J/kg/K (typical specific heat for steels)
   $T_{\text{av}} = 3 \times 10^3$ K (value near to boiling point)

   Applying these values gives a predicted weld width of 2 mm, which lies within the observed range for mild steels at these powers.

2.10 THE WORKPIECE INTERACTION

Having discussed heat flow processes within the workpiece, this section describes the physical bases of laser beam absorption both at surfaces and within hot gaseous media, and presents possible mechanisms for fluid flow within the molten zone.

2.10.1 Maxwell's equations

The propagation of electromagnetic waves (light) is based on four classical equations:

Gauss' theorem applied to electrostatics $\text{div}\ D = \rho$ (2.11)

where div is a vector operator, D is the electric displacement and ρ is the electric charge density;

Gauss' theorem applied to magnetostatics $\text{div}\ B = 0$ (2.12)

where B is the magnetic induction;

Faraday's and Lenz's laws of induction $\text{curl}\ E = \partial B/\partial t$ (2.13)

where curl is a vector operator, E is the electric field and t is time;

Ampère's law modified by Maxwell $\text{curl}\ H = \sigma E + \partial D/\partial t$ (2.14)

where σ is the conductivity of the material.

Taken together, these equations, known as Maxwell's Equations, accurately predict the behaviour of light waves in terms of propagation through transparent and absorbing materials. For further information the interested reader is referred to any undergraduate text on electricity and magnetism.

2.10.2 Absorption at a surface

Equation (2.14) can be rewritten as

$$\text{curl}\ H = \varepsilon\varepsilon_0(\partial E/\partial t) + \sigma E \tag{2.15}$$

The first term on the right-hand side represents displacement current and describes light travelling through insulators such as lenses, windows and prisms. Propagation by this means alone is lossless. The second term represents conduction current and describes light travelling through conductors such as metals, and bouncing of the surfaces of mirrors. Propagation by this means is lossy. For a laser beam of single frequency ν, Maxwell's equations reduce to

$$(n/c)E = \mu\mu_0 H \tag{2.16}$$

and

$$(n/c)H = (2\pi\nu\varepsilon\varepsilon_0 - \text{j}\sigma)E \tag{2.17}$$

where n is the refractive index, c the velocity of light and j is a unit vector.

If the laser frequency ν is less than $\sigma/2\pi\nu\varepsilon\varepsilon_0$, then the conduction current term in equation (2.17) dominates and the laser beam is heavily absorbed. For a typical metal this occurs for frequencies less than about 10^{15} Hz, which includes in its range both YAG and CO_2 lasers. The laser beam is absorbed in a distance

$$\partial = (\pi\sigma\nu\mu\mu_0)^{-1/2} \tag{2.18}$$

called the skin depth. For the YAG laser the skin depth of a typical steel is about 10 nm, and for the CO_2 laser about 30 nm. The absorbed energy is thus deposited within the first few atomic layers of the material surface. Energy not directly absorbed will be reflected with a coefficient

$$R = 1 - 4/\mu_0 \sigma c \partial \qquad (2.19)$$

For a clean, cold metal surface, R for a YAG laser beam is about 88% according to this formula, and 96% for a CO_2 laser. In practice both figures will be less than those quoted due to electronic transitions, other absorption processes, surface irregularities and contamination.

2.10.3 Absorption in a hot gaseous medium

From an analysis similar to the above, it can be shown that very hot metal vapours (plasmas) also exhibit a critical frequency. It is called the plasma frequency and is given by

$$\nu_{\text{plasma}} = (ne^2/4\pi\varepsilon_0 m)^{1/2} \qquad (2.20)$$

where n and e are the electron density and charge respectively. Numerically this has the value $9n^{1/2}$ Hz (n is number per cubic metre). For laser beams below this frequency, the plasma reacts rapidly to electric fields and the laser beam is scattered almost without loss. This is the origin of the so-called plasma effect. Above this frequency the beam can propagate through the gas and be absorbed over a large distance by various collision processes both within the gas and its surroundings.

2.11 EVOLUTION OF THE WELD ZONE

2.11.1 Stage 1

The intensely focused laser beam ($\approx 10^4$ W/mm^2) impinges on the metal surface. Most of the beam is initially reflected, equation (2.19), but part is absorbed into the top few atomic layers, equation (2.18). Heat is conducted away into the bulk of the metal setting up a time-dependent hemispherical temperature profile in the solid. As the metal heats up, its resistivity increases, and hence its absorptivity, as defined by equation (2.19), increases, Fig. 2.13.

2.11.2 Stage 2

For low powers, or defocused beams, the metal will melt but not vaporize, giving rise to a roughly hemispherical molten zone, equations (2.4, 2.5), and Fig. 2.14. In practice, convection within the weld zone, and other factors, will distort the shape away from hemispherical. This type of weld is known as conduction-limited, and is characterized by a width between

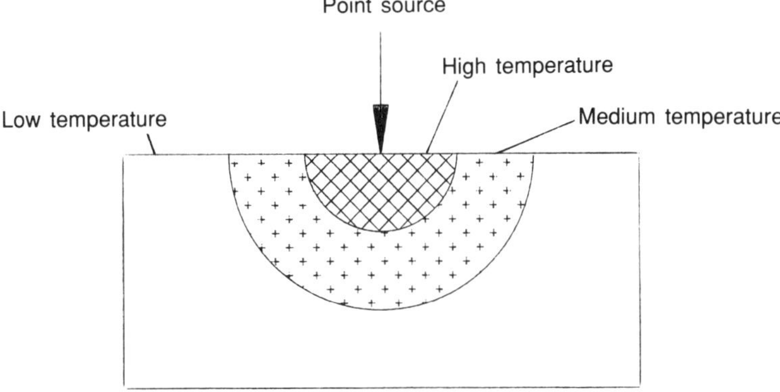

Fig. 2.13 Temperature profile for stationary point source.

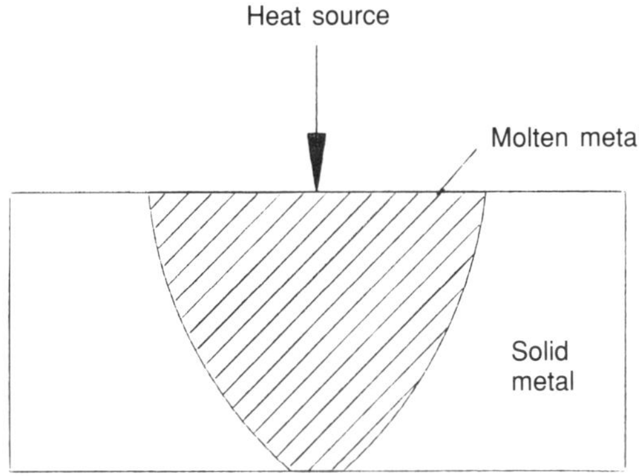

Fig. 2.14 Conduction limited weld zone.

one and two times the depth. Similar shape welds are made by most arc welding processes.

2.11.3 Stage 3

At higher powers and near to focus, the metal will boil, equations (2.8, 2.9), and three related events taken place:

1. Vapour is liberated which depresses the liquid surface due to reaction pressure. The depressed surface improves the beam absorption by forming an absorbing cavity.
2. The vapour becomes ionized and further beam absorption takes place directly through Inverse Bremsstrahlung (collisions between photons and free electrons), and indirectly via absorption in the liquid metal surrounding the depression.

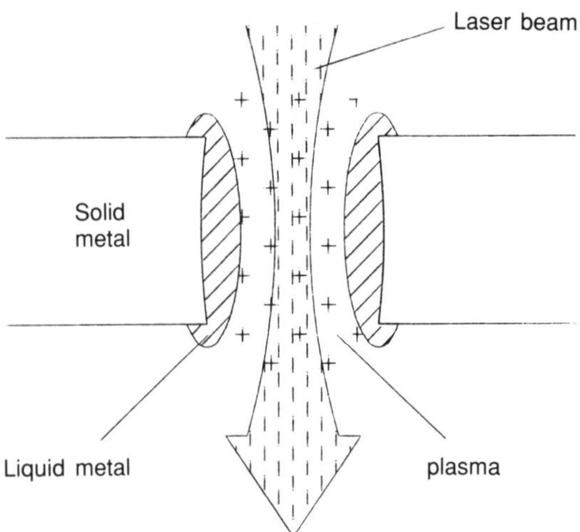

Fig. 2.15 Cross-section of keyhole.

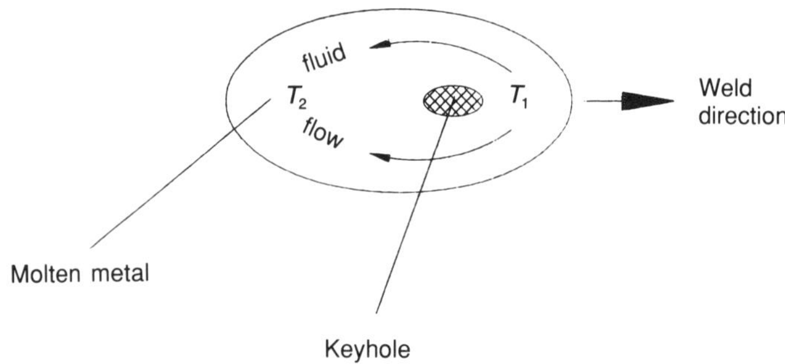

Fig. 2.16 Fluid flow in weld due to surface tension.

3. The depression penetrates completely through the material forming a deep penetration keyhole, Fig. 2.15.

2.11.4 Motion of the weld zone

For conduction-limited welds, and the exposed extremities of deep penetration keyholes, a possible mechanism for fluid flow is surface tension imbalance. Fig. 2.16 shows the weld zone, viewed from the direction of the incoming laser beam. At the point of impact, the metal will be at the boiling point. Further away, the temperature will be lower. At temperatures close to the boiling point, surface tension decreases as temperature rises. Thus the region marked T_1 will be hotter than the region marked T_2

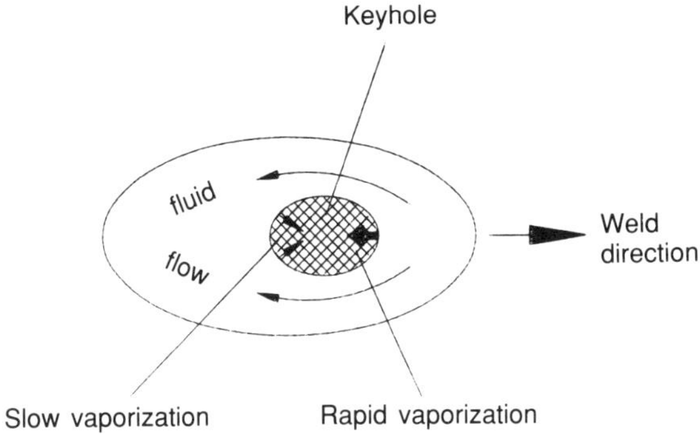

Fig. 2.17 Fluid flow in weld due to differential vaporization.

and there will be a net surface tension force from T_1 to T_2 with a corresponding metal flow in the same direction.

Within a deep penetration keyhole, the exposed liquid surfaces are all likely to be at or near to the boiling point, and thus surface tension imbalance will be minimal. The probable cause of fluid flow in this case is due to the difference in evaporation rates. The liquid surface nearer to the beam will evaporate at a higher rate than the surface behind the beam, and the difference in reaction pressure will cause fluid flow from front to back of the keyhole, Fig. 2.17.

The above explanations for fluid flow are simplistic and approximate, yet give a considerable degree of physical insight into the mechanisms involved. Readers interested in a more rigorous approach are referred to the references, and in particular to Dowden *et al.* (1985a, 1985b).

2.12 ACKNOWLEDGEMENTS

The authors are indebted to Mr M. Amin for helpful and stimulating discussions, to Miss N.A. Hall and Mrs P.S. Edmundson for preparation of the manuscript, to Mr P.D. Evans for assistance with the illustrations, and to TWI for permission to publish this material.

REFERENCES

Anon. (1979) *Weld. J.* **58**(7), July, 53–4.

Ayers, J.D., Tucker, T.R. and Schaeffer, R.J. (1981) *Source Book on Applications of the Laser in Metalworking*, ed. Metzbower, E.A., Metals Park OH44037, USA, American Society for Metals, 301–9.

Banas, C.M. (1989) US Patent NO. 4152575, May.

Breinan, E.M. and Banas, C.M. (1974) *WRC Bulletin*, *201*, December, 47–57.

Breinan, E.M., Thompson, E.R., Banas C.M. and Kerr, B.H. (1977) *United Technology Corporation Research Centre Report R77-9122887–3*.

Carslaw, H.S. and Jaeger, J.C. (1959) *Conduction of Heat in Solids*, 2nd ed., Oxford University Press.
Courtney, C. and Steen, W.M. (1978) *Applied Physics* (Germany), **17**(3), November, 303–7.
Crafer, R.C. and Oakley, P.J. (1981) *Welding Institute Research Report 165/1981*, October.
Dawes, C.J. (1983) *Proc. Conf. Developments and Innovations for Improved Welding Production*, Birmingham, September.
Dowden, J., Davis, M. and Kapadia, P. (1985a) *Journal of Applied Physics*, **57**, 4474–9.
Ibid. (1985b) *Journal of Physics D*, **18**, 1987–94.
Eckersley, J.S. (1982) *Proc. Conf. ICALEO 1982*, Boston, September.
Estes, C.L. and Turner, P.W. (1974) *Weld. J.*, **53**(2), February, 662–73s.
Foulk, L.R. (1978) *Report DBX-613-1841 (Rev)*, Kansas City, MO, Bendix Corp, November, 32.
Fraser, F.W. and Metzbower, E.A. (1983) *Proc. Conf. Applications of Lasers in Material Processing*, January, Los Angeles.
Gnanamuthu, D.S. (1979) *Proc. Conf. Applications of Lasers in Materials Processing*, Washington, DC, April, 177–211.
Goldak, J.A. and Nguyen, D.A. (1977) *Weld. J.*, **56**(4), April, 119s-125s.
Grosjean, D.F., Olson, R.A. and Sarka, B., Jr. (1978) *Rev. Sci. Inst.* (USA), **49**(6), June, 778–81.
Harker, K.J. (1975) Use of scanning slits for obtaining the current distribution in electron beams. *J. Applied Physics*, **28**(11), 1354–7.
Hella, R.A. (1978) *Optical Engineering*, **17**(3), 198–201.
Jimbou, R. *et al.* (1980) *Proc. Int. Conf. Welding Research in the Eighties*, Japanese Welding Research Institute, Osaka.
Lim, G.C. and Steen, W.M. (1982) *Optics and Laser Technology*, **14**(3), June, 149–53.
Meyerhofer, D. (1968) *IEEE Journal of Quantum Electronics*, **4**(11), November, pp. 969–70.
Moon, D.W. and Metzbower, E.A. (1983) *Weld. J.*, **62**(2), February, 53s-58s.
Pauley, J.T. and Russel, J.D. (1978) US Patent No. 4127761, November.
Saifi, M.A. and Vahavioios, S.J. (1976) *IEEE J. of Quantum Electronics*, QE-**12**(92), February, 129–36.
Sanderson, A. (1975) *Welding Institute Research Report Misc 34*, October 1975.
Sepold, G., Juptner, W. and Rothe, R. (1980) *Proc. Int. Conf. Welding Research in the 1980s*, Osaka, October, Addition paper A-29, 3.
Seretsky, J. and Ryba, E.R. (1976) *Weld. J.*, **55**(7), July, 208s–211s.
Shinada, K. *et al.* (1981) *Proc. Int. Conf. Laser Processing,* Anaheim, Ca, USA, November, paper 3.
Stoop, J. and Metzbower, E.A. (1978) *Weld. J.*, **57**(11), November, 345s-353s.
Willigoss, R.A. *et al.* (1979) *Weld Metal Fab.*, **47**(2), March, 117–27.

3

Beam transmission systems

D.E. Greening

3.1 INTRODUCTION

In one chapter it is not possible to cover the details and requirements for a wide combination of beam delivery systems and laser types. Here an attempt is made to approach some general considerations of beam transmission systems, without resorting to the detailed theory of Gaussian optics which describes the propagation of laser beams. Reference is mainly to CO_2 laser systems, the type most commonly used for industrial materials processing. Many of the topics covered will have general application to other types of laser. The main approach will be through optical considerations, since other aspects involving safety, nozzles, assist gases, etc. are covered in other chapters.

We can regard the topic of beam transmission systems as referring to delivery of the laser beam from the laser to the workpiece, and discuss the design, construction and operation of the system from the viewpoint of extracting the best operational efficiency from a given laser. Aspects of laser beam delivery to be discussed are shown in Fig. 3.1. It is, to some extent, artificial to break down the topic of beam delivery systems into self-contained topics, since many are interrelated. For example, operator safety generally involves complete 'piping-in' of the beam path. This has implications for both the rigidity of the optical system (and hence the ability to maintain accurate alignment) and for cleanliness, since a reasonably sealed system allows the introduction of some flowing gas to keep contaminating particles away from critical optical surfaces.

3.2 MECHANICAL DESIGN CONSIDERATIONS

3.2.1 Safety

For our purposes it is sufficient to note that all sections of the beam path should be enclosed by piping of a type adequate to provide system rigidity,

Laser Processing in Manufacturing. Edited by R.C. Crater and P.J. Oakley.
Published in 1993 by Chapman & Hall, London. ISBN 0 412 41520 8

MECHANICAL DESIGN
Safety/'Piping-in'
Cleanliness/Avoiding damage to optics
Reduction of back spatter
Nozzles/Assist gases

OPTICAL DESIGN
Beam properties/Laser modes/Beam clipping
'Q' Factor/Modified gaussian optics
Focal spot sizes/Power density/Depth of focus
Beam expanders/Reduction of divergence

Fig. 3.1 System considerations.

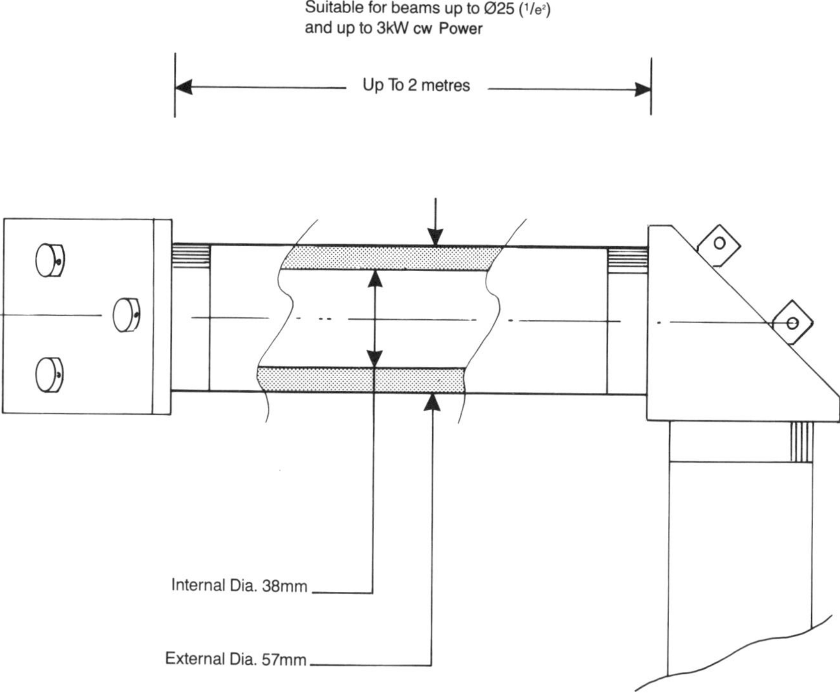

Fig. 3.2a Typical piping.

and proof against accidental misalignment leading to the laser beam striking the piping. In certain cases sealing may not be possible – perhaps where a sub-assembly may need lateral adjustment. Figure 3.2a indicates typical piping for CO_2 lasers of powers around 1–3 kW cw; and Fig. 3.2b shows an example of a beam expander requiring lateral adjustment. In this latter case a 'shield-pipe' can be designed to avoid any possibility of the

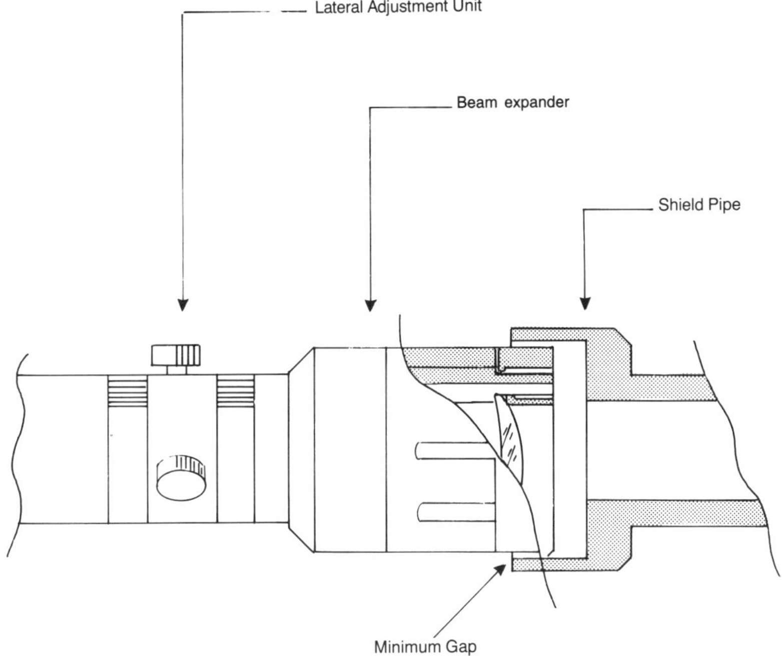

Fig. 3.2b Shielding a sub-assembly.

system operator's fingers interrupting the beam during adjustment of the beam expander.

3.2.2 Cleanliness and avoiding damage to optics

Contamination of the optical surfaces of mirrors and lenses by grease (fingerprints, etc.) or dust can dramatically reduce the damage threshold of the components. (Ideally, the optics should be cleaned using a soft lint-free lens cloth and water-free acetone.) Between the laser output window and the focusing element there may be several other optical surfaces, either reflective or transmissive. In a fully-piped system it is often worthwhile to introduce some positive gas pressure to keep out dust which may otherwise land on these surfaces. In high-power laser systems particles of dust can be burned into the optical surface, causing failure due to localized high absorption.

One area where cleanliness cannot always be maintained is the final optical surface before the workpiece, which may suffer from bombardment by back-spattered material ejected from the workpiece. Figure 3.3 shows a typical focusing lens arrangement. The amount of back-spatter reaching the lens can be reduced in several ways. Firstly, a flow of assist gas through

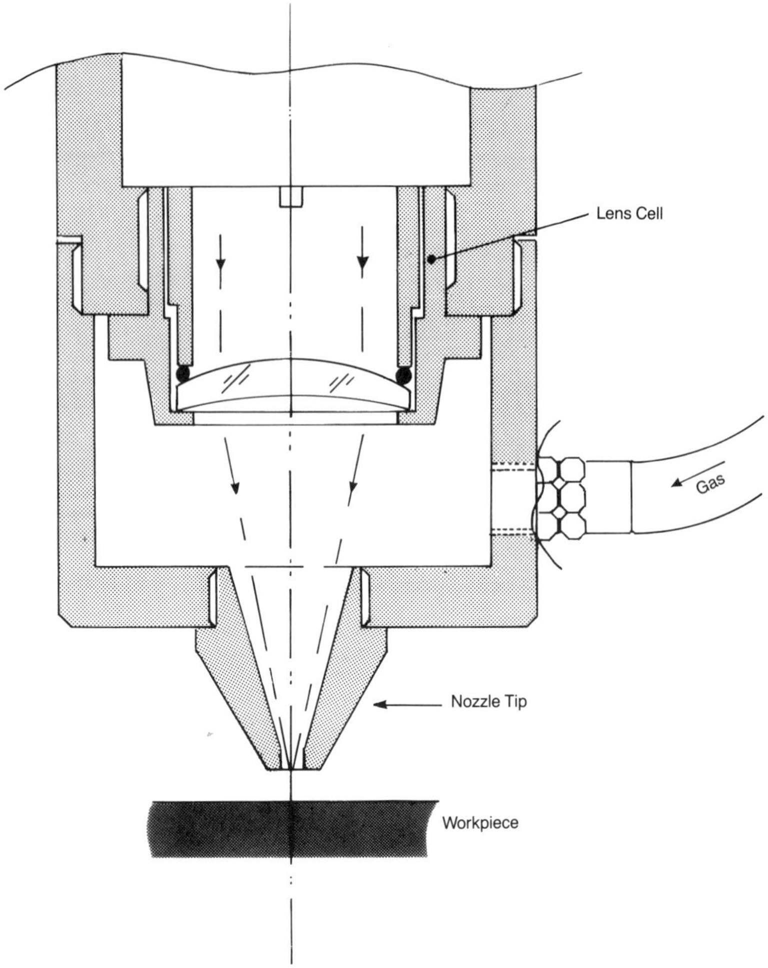

Fig. 3.3 Typical nozzle (schematic).

the nozzle will reduce the amount of back-spatter reaching the lens. Secondly, an air flow can be introduced across the lens surface to deflect the back-spattered material. Thirdly, protective windows can be used. Unfortunately, in CO_2 laser systems any coated window is likely to be nearly as expensive as the lens. Uncoated KCl windows can be used but these typically cause a loss of around 10% of transmitted power due to back-reflections. A fourth approach is described in section 3.4.5.

During the design of the system a choice must be made whether to use transmissive or reflective optics to focus the beam. In CO_2 laser systems transmissive optics (ZnSe) can be used up to about 5 kW total power, and

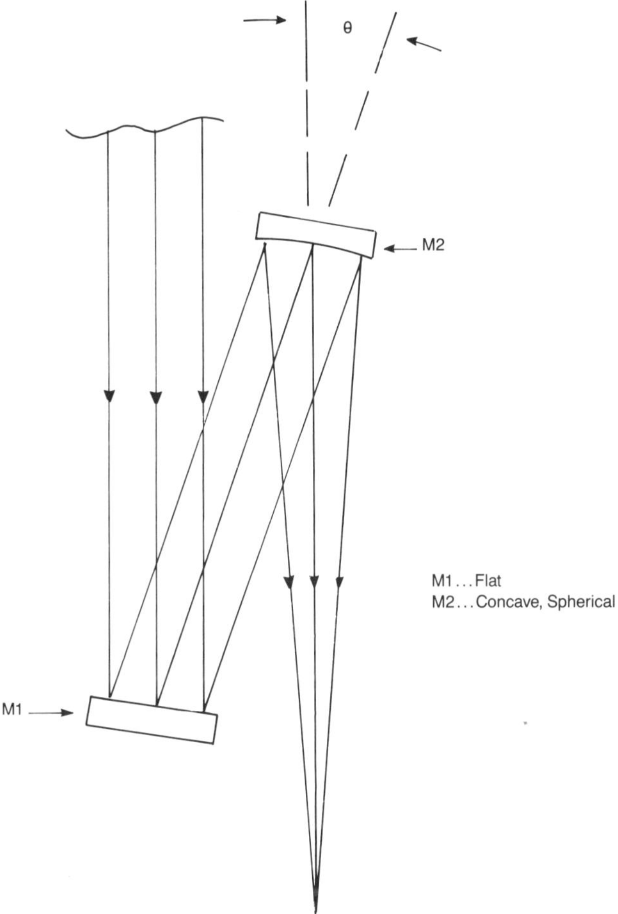

Fig. 3.4a 'Z' configuration.

at power densities of around 1 kW/cm^2, where reasonable cleanliness can be maintained. Pulsed CO_2 laser damage thresholds of around 100–200 MW/cm^2 are typical, but are limited by optical coatings rather than the bulk material of the component. Reflective focusing systems will withstand higher power levels and are more resistant to back-spatter.

Figure 3.4a shows a Z-configuration mirror system such as may be used in high-power welding operations. Off-axis aberrations are minimized by keeping the angle θ as small as is practicable, but the relatively long focal length of the mirror leads in any case to a large focused spot size. Figure 3.4b shows a short-focal-length mirror system using a non-spherical mirror

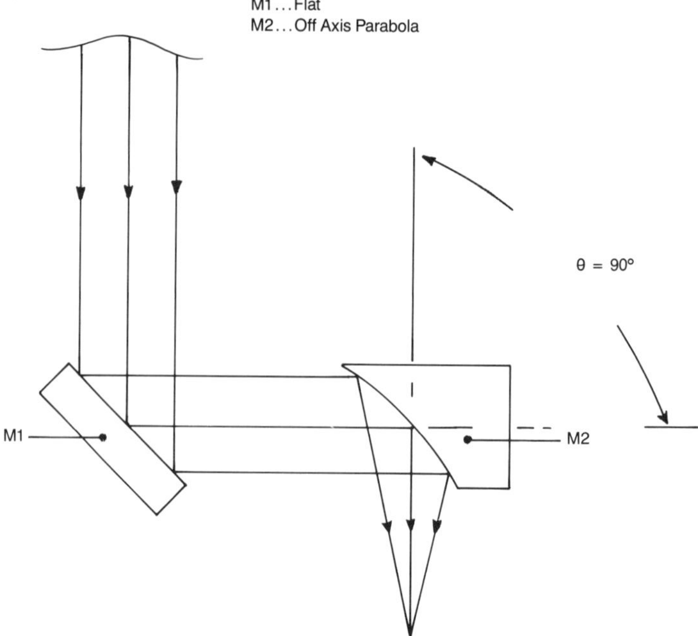

Fig. 3.4b Short focal length mirror system.

to produce a small, aberration-reduced focused spot. Such aspheric mirrors are generally produced by diamond machining, and can be very expensive. The optical format described in section 3.4.5 could equally be used to protect such mirrors from excessive back-spatter.

3.2.3 Nozzles and assist gases

Nozzle design, and specification of assist gas types and pressures, can be considered to be part of the materials processing aspect of a laser installation. We will not cover any of this area in detail, except to make the following two observations: 1. Even where the optical performance of a laser system is improved, in terms of power density/depth of focus, the system performance may *not* be improved unless suitable changes are made in the nozzle and gas flow areas. 2. Where high gas pressures are envisaged, in cutting certain metals such as aluminium for example, the focusing lens or sealing window which must withstand this excess pressure should be designed accordingly.

Appendix C gives a 'rule-of-thumb' formula for this calculation.

(COMPUTER GENERATED FIGURES)
Beam Dia. (1/e²) vs. Distance

DIST. FROM LASER	TEMoo	'REAL' BEAM
0	8.72	13.08
1m	8.86	13.28
2m	9.25	14.82
3m	9.88	16.04
4m	10.69	17.49
5m	11.66	19.11
6m	12.74	20.86
7m	13.91	22.72
8m	15.14	24.65
9m	16.43	26.65
10m	17.76	28.69

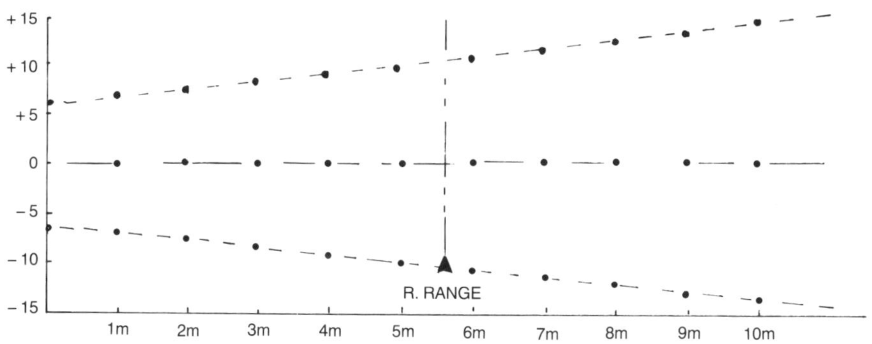

Fig. 3.5 Propagation of CO_2 laser beams.

3.3 OPTICAL CONSIDERATIONS

3.3.1 Lateral modes

No laser beam can be focused to an indefinitely small point, and no laser beam can be made to be exactly parallel as it propagates through space. There is a wide range of textbooks in which the equations describing the propagation of laser beams are stated. These 'Gaussian Optics' equations usually refer to the properties of an ideal beam. Real lasers tend to emit a beam which is a superposition of 'modes', that is, ways in which the resonant cavity can lase. The narrowest, least divergent, of these modes is labelled 'TEM_{00}' and has the properties described by the Gaussian equations mentioned above.

Figure 3.5 takes the case of an existing laser cavity design and demonstrates the calculated output properties of an ideal, TEM_{00}, beam, and of a typical real output from that cavity. The cause of the wider beam is due to emission from the cavity of higher-order modes. In the case shown, a large amount of the laser energy is emitted in the next-higher mode, known as 'TEM_{01*}'. At all points in space this latter beam is about 50% wider than the TEM_{00} mode, and so is 50% more divergent. If even higher order modes were allowed to lase in the cavity, the beam diameter and divergence would be even greater.

This effect can sometimes be observed in real industrial CO_2 lasers, where the beam diameter and divergence increase as the plasma current is increased in the laser cavity and the laser emits more power. The relative fraction of higher-order modes increases, leading to reduced beam quality.

Some particular points concerning Fig. 3.5 are:

1. The divergence of the beam increases with distance from the beam waist, asymptotically reaching a maximum value at infinity. The distance at which the divergence reaches $1/\sqrt{2}$ of its maximum value is known as the Rayleigh range, about 5 m in the case shown in Fig. 3.5, and this is also the position at which the beam has grown to a size $\sqrt{2}$ times larger than its diameter at the beam waist.
2. The eventual divergence, at infinity, is known as the far-field divergence, and is around 2.3 mrad in the example shown in Fig. 3.5.
3. The cavity in this case is designed to have a beam waist at the output window. For the case of the TEM_{00} beam, the diameter at the waist (in mm) multiplied by the far-field divergence (in milliradians) always gives the number 13.5, for a CO_2 laser. The derivation of this number is given in Appendix A.

Figure 3.6 shows a sketch of the intensity profiles for TEM_{00} and TEM_{01*} beams independently, and indicates the type of intensity profiles of a real beam consisting of a mixture of these two modes. Note that the beam diameter for calculation purposes is measured at the $1/e^2$ points of intensity (about 13.5% of peak intensity), and that some beam energy exists outside this diameter. This must be considered when designing the optical clear aperture(s), or power will be lost due to 'beam clipping'.

3.3.2 'Q' factor

Laser beams with higher order lateral modes (TEM_{01}, TEM_{21}, etc.) have, everywhere in space, a greater diameter and divergence than a TEM_{00} mode. For a variety of reasons, it is useful to establish a measure of 'quality' of the output of real lasers. Such a measure can be called the 'Q' factor. Referring back to the computer-generated figures in Fig. 3.5, the TEM_{00} mode has a far-field divergence of 1.548 mrad. Multiplying the beam diameter near the laser (at a waist) by the far-field divergence we obtain the figure: $8.72 \times 1.548 = 13.5$. However the cavity design is changed,

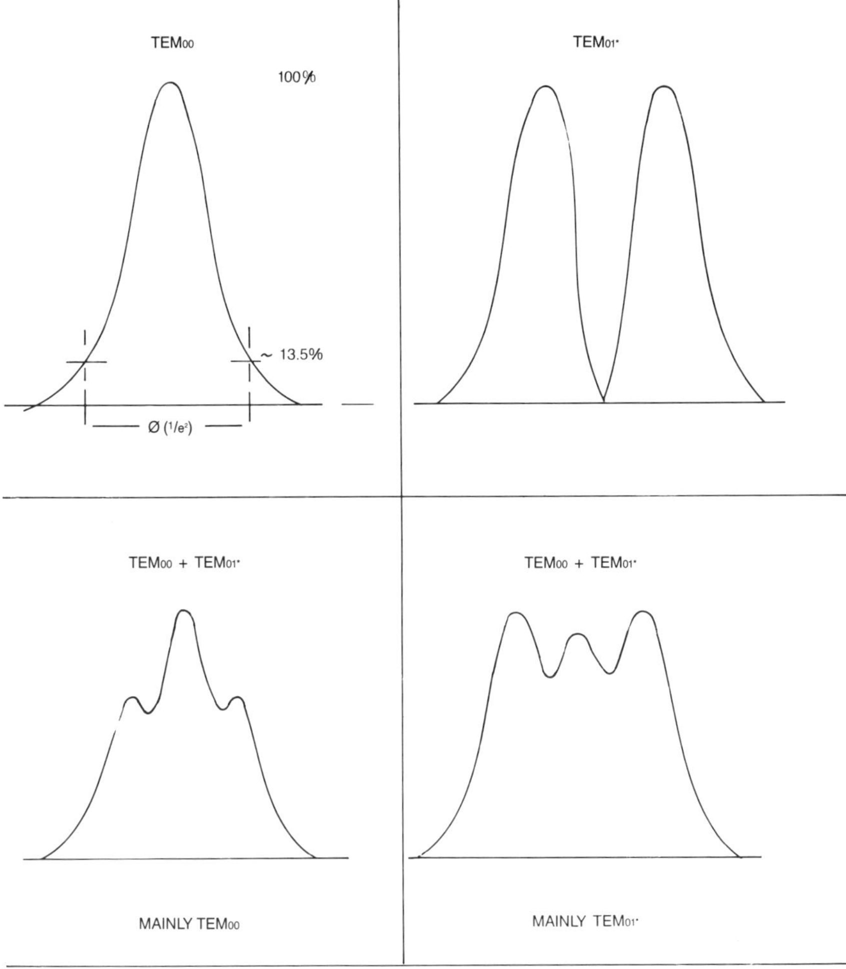

Fig. 3.6 Intensity profiles (schematic).

these figures will always multiply to this value, for a CO_2 laser for the TEM_{00} mode. Appendix A gives the reason for this.

A real laser will always emit a beam where the waist diameter D ($1/e^2$) in mm multiplied by the far-field divergence Vm gives a value **larger** than 13.5. The computer-generated figures of Fig. 3.5 were based on an output of a (mainly) TEM_{01*} beam mode. For this beam $D.Vm = 13.08 \times 2.32$ mrad $= 30.35$.

$$\text{For a real beam we define } Q = D.Vm/13.5 \qquad (3.1)$$

In the above case, $Q = 30.35/13.5 = 2.25$.

(*Note*: Some publications, notably these from the company Coherent Inc., use a value M^2 to express a measure of beam quality. This M^2 value is identical to our Q value.)

Use of the Q factor enables the textbook Gaussian equations to be modified for real laser systems. Wherever the value λ for wavelength appears in the equations, substitution of the value $Q\lambda$ will give the correct result. An example of this is as follows:

The limit of resolution of a laser beam is given by the Gaussian equation for spot size = (X mm).

$$\begin{aligned} X &= 2\lambda/\pi\text{NA} \qquad (3.2) \\ &= 0.0106 \text{ mm for a } CO_2 \text{ laser} \end{aligned}$$

where NA = 'used' lens numerical aperture.

For the sake of argument, let us chose NA = 0.25, F/no = 2.0. Suppose we put a perfect lens of 26.16 mm focal length in the real beam of Fig. 3.5. The actual spot size formed will (by definition of far-field divergence), be 26.16 × 2.32/1000 = 60.7 μm in size. See equation (3.4) later.

The Gaussian equation predicts (NA = 0.25)

$$\begin{aligned} X = 0.0106 \times 0.637/0.25 &= 0.027 \text{ mm} \qquad (3.3a) \\ &= 27\ \mu\text{m} \end{aligned}$$

The modified equation, using $Q\lambda$ instead of λ

$$X = 2Q\lambda/\pi\text{NA} \qquad (3.3b)$$

So, X = 60.7 μm which is consistent with the discussion above.

In the following section we will see that the optical performance of a system is critically dependent upon the laser Q factor.

3.3.3 At the focus

Two factors contribute to the focused spot size for a laser beam. The first of these is the divergence contribution considered earlier, namely:

$$X \text{ mm} = \text{lens focal length (mm)} \times \text{far-field divergence in radians, or} \qquad (3.4a)$$

$$X\ \mu\text{m} = \text{lens focal length (mm)} \times \text{far-field divergence in milliradians} \qquad (3.4b)$$

The second contribution to spot size is given by the aberrations of the focusing element. We will consider only the on-axis type known as 'spherical aberration'. Off-axis types: coma and astigmatism occur when using spherical mirrors in Z-configurations, but are difficult to discuss in brief. Optical designers using computer raytracing techniques should be involved if such systems are contemplated. Taking the example of focusing lenses,

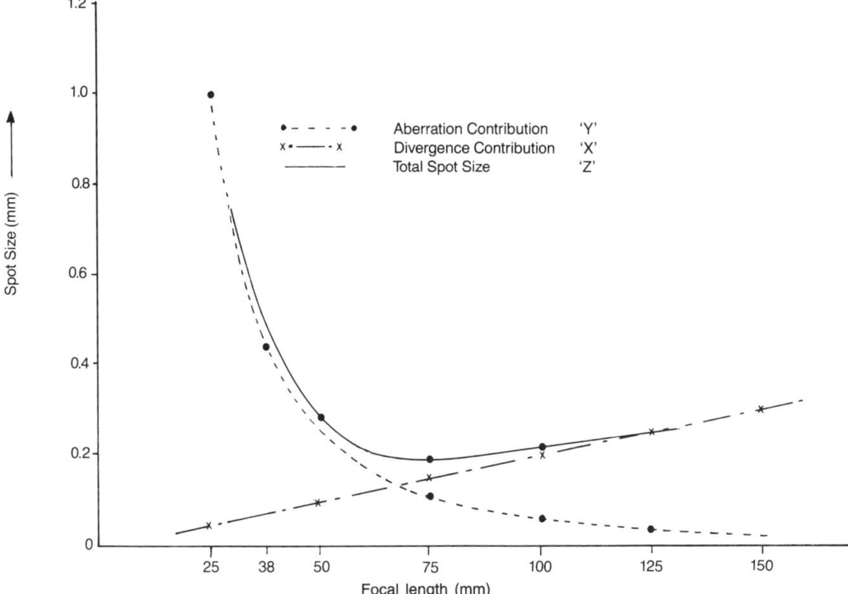

Fig. 3.7 Summation of contributions.

it is possible to obtain a good approximation to the aberration contribution Y (mm) by a simple formula, assuming we are focusing a near-parallel beam.

$$Y = kd^3/F^2 \tag{3.5}$$

where

d = beam diameter at the lens,
f = lens focal length,
k = a constant.

For a ZnSe lens, the constant k is roughly $k = 0.04$ for a meniscus lens and $k = 0.06$ for a plano-convex lens.

The local spot size is approximated by

$$Z \text{ (total spot size)} = \sqrt{(X^2 + Y^2)} \tag{3.6}$$

Figure 3.7 shows how the total spot size varies with focal length for a laser beam of 25 mm diameter, and 2 mrad far-field divergence. The lens is assumed to be a ZnSe meniscus type. Where high-energy densities (small spot sizes) are desirable, in processes such as ceramics scribing, it is beneficial to use focusing lenses of an 'aberration-free' type. These may be, for example two-element corrected lenses or lenses with aspheric surfaces.

Table 3.1 compares a single-element ZnSe meniscus lens of 38 mm focal length and an equivalent two-element lens of 38 mm focal length in a

Table 3.1 Performance comparison of single- and two-element lenses

	Spot size (μm)	Average power density (kW/mm^2)
Meniscus lens	227.0	3.7
Corrected lens	49.4	76.4

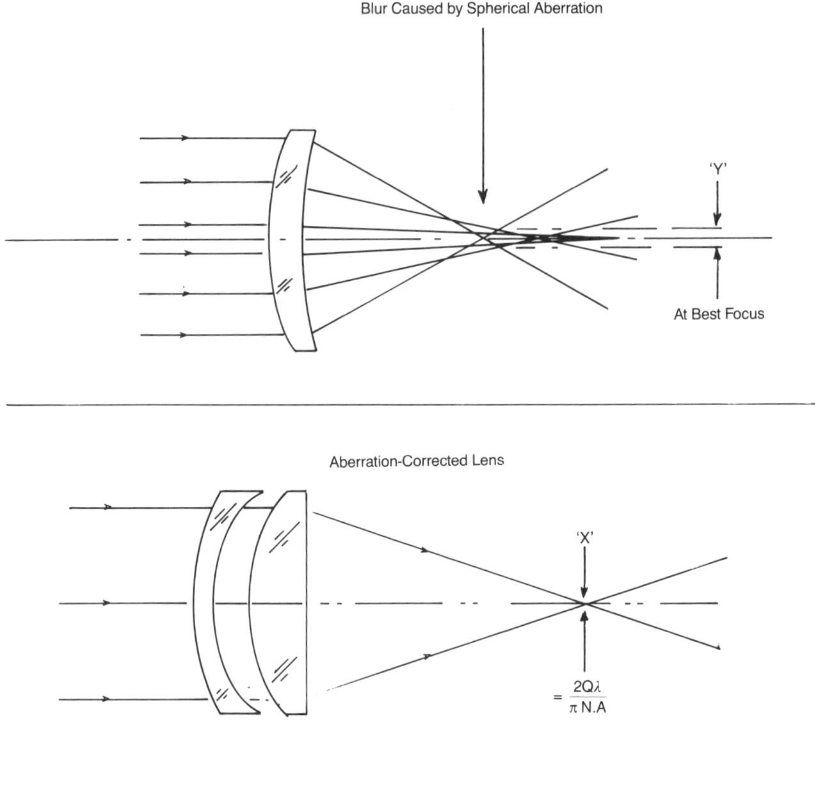

Fig. 3.8 Contributions to spot size.

ceramics-scribing application with a laser beam diameter of 20 mm and a far-field divergence of 1.3 mrad, and average power 150 watts (Fig. 3.8).

A major factor influencing system performance is depth of focus. In real systems this is the range of focus positions over which the process can be said to work reasonably well. Mathematically it is an arbitrary range which can be derived, using the Gaussian equations, and which depends on the allowed variation in spot size. A typical definition is the range over which

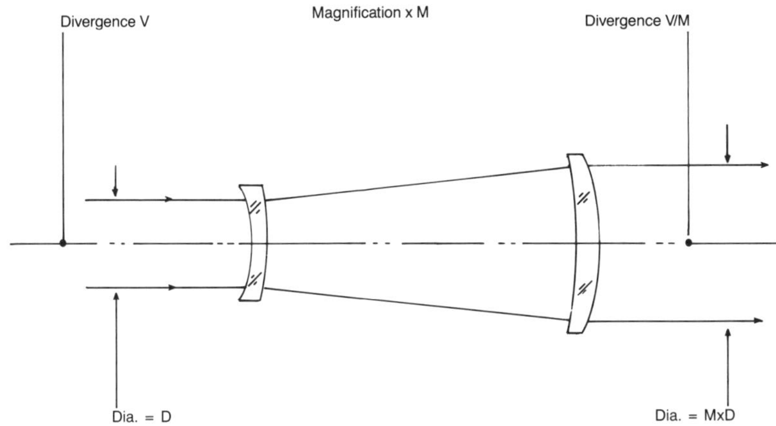

Fig. 3.9 Galilean beam expander.

the focused spot increases, either side of the best focus, to 105% of its minimum value. Real processes may be more, or less, sensitive than this.

Laser beams can be manipulated by the optical system to give optimum results for a given set of circumstances. Appendix D shows that, with aberrations eliminated, the following facts are invariably true, for a given laser:

1. For a fixed *F*/number, power density is proportional to $1/Q^2$.
2. For a fixed Q factor, depth of focus is proportional to 1/(power density).

3.3.4 Beam expanders

If a laser beam is expanded by a factor M, its far-field divergence is reduced by that factor. Figure 3.9 shows a typical beam expander format, the negative and positive lenses forming a kind of 'reverse' telescope. This is called a Galilean beam expander. There are two main classes of use for such instruments. Firstly, small fixed-focus devices can be used in certain systems to reduce the beam divergence immediately prior to an aberration-corrected focusing lens.

For a magnification M:

Spot size decreases by a factor M,
Power density increases by a factor M^2,
Depth of focus decreases by a factor $1/M^2$.

Typically such arrangements are used in the scribing of ceramics for large-scale integrated circuits.

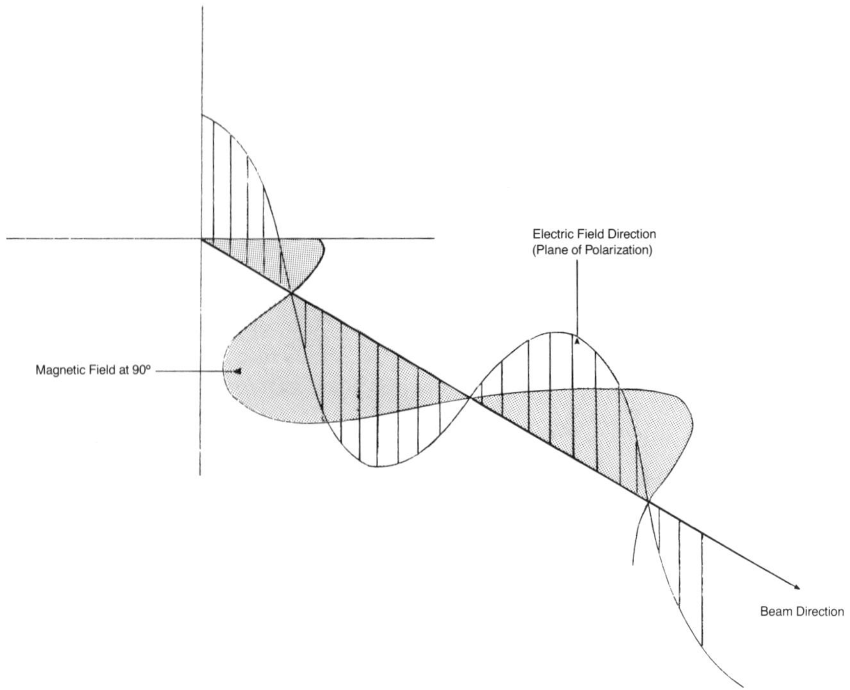

Fig. 3.10 Plane of polarization.

Secondly, a focusable beam expander may be placed near the laser to control the beam in a long-path, 'moving-optics' system. Wherever, for example, a set of mirrors is used to direct a beam over a large cutting bed, several metres in x and y directions, a beam expander may be used to maintain a constant power density and depth of focus over the whole bed. Appendix B gives the details of one such application. Ill-considered use of beam expanders can degrade the optical system where single-element lenses are used. Appropriate use should be made of equations (3.4) and (3.5) in such a case!

3.3.5 Polarization

Laser beams are oscillating electric and magnetic fields propagating through space. The direction of the electric field, and its strength, can be described by the electric vector for that field, and the direction of the vector, (in terms of azimuth angle, or angle of 'rotation'), described as the state of polarization (Fig. 3.10).

Lasers can be constructed such that the state of polarization fluctuates quickly and randomly. Such a beam is said to be randomly polarized.

However, many lasers are built to reinforce the lasing action when the electric field is in a specific orientation. The emitted beam then maintains a strong polarization in one specific orientation. This is linear polarization.

The importance of polarization is apparent when processing metals. For example, in cutting metal with a laser, if the relative direction of movement of the metal to the focused spot is in the direction of polarization, the metal will cut easily. Across the polarization the metal cuts less well, in extreme cases either not penetrating the metal or leading to unacceptable finish. A generally accepted way of overcoming this problem is the installation of an optical component or device known as a phase retarder.

The effect of a phase retarder is to spin the direction of polarization at the frequency of the radiation (about 30 THz for a CO_2 laser), and to even-out the polarization effects during processing. This state of polarization, with the electric field vector spinning once every wavelength, is known as circular polarization. Some laser manufacturers install phase retarders within their laser units. In other cases due consideration must be given to the design of the beam delivery system if a phase retarder is required, since they operate correctly only at compound angles of 45° incidence and 45° azimuth.

In-line phase retarder units are available, avoiding dog-legs in the delivery piping.

3.3.5 Energy transfer and alignment

In order to relay the maximum emitted laser energy to the workpiece the correct optical elements should be chosen. Transmissive components for CO_2 lasers usually present no problem, since modern coatings allow ZnSe lenses to be made with total absorption levels of around 0.1% and surface reflections of about the same value, giving an overall transmission of about 99.7%, or better, per component. Mirrors in these systems present a set of problems where a trade-off of reflectivity/lifetime/cost must be made. Table 3.2 gives the details of some types of CO_2 laser mirrors.

A second consideration concerns the optical clear aperture. At any position in the system it is desirable to have optical components with a clear aperture about 30% larger than the $(1/e^2)$ beam diameter, in order to avoid clipping the extremes of the energy distribution. Alignment of the laser beam throughout the system is also of importance, since a beam passing non-centrally can be 'clipped' by mechanical stops in the optical path. Of course, misalignment can also degrade optical performance in other ways, creating non-symmetrical distributions of energy in the focused spot.

Where a visible sighting beam, from a HeNe laser for example, is not available, the standard method of alignment involves the use of a copper crosswire. The crosswire is placed, mechanically centred, at some location

Table 3.2 Comparison of some CO_2 laser mirror types

Mirror type	*R*%	Comment
Nickel-copper (Gold coated)	98.6–99.0	Cheap, available, low phase shift, medium durability
Enhanced NiCu	99.8	Approx. 2 times price of NiCu, less available in all types
Silicon	99.7	Price depends on quantity, can cause phase shifts up to 10°
Silicon zero phase shift	99.7	Expensive, currently from USA only
Molybdenum	98.5	Approximately $2^1/2$ times cost of NiCu, **very** durable
Cu Diamond m/c (Gold coated)	98.6–99.0	No significant advantage unless aspheric form required

in the delivery system, and the laser is fired briefly at a target fixed just beyond the crosswire. The shadow of the crosswire on the resulting burn pattern indicates the degree of centration of the beam, which is corrected if in error. This procedure is repeated at other locations in the delivery system until the beam is passing centrally at all locations.

3.4 CURRENT AND NEAR-FUTURE DEVELOPMENTS

3.4.1 Mode forcing

Special graded reflectivity coatings have been applied to the output coupler in a variety of lasers in order to attempt or force the laser to emit the TEM_{00} mode only. Although successful in certain laser types, in high-power CO_2 lasers these special coatings are reported as having too high an absorption level, leading to thermal lensing. Such coatings are very expensive. V & S Scientific have recently developed a device to replicate the effects of a graded coating, but using interferometric principles. No results are available to date, as final tests have yet to take place. If successful, CO_2 lasers of a variety of makes, and up to at least 2 kW c.w. power, can be forced to emit the TEM_{00} mode (giving $Q = 1$), with negligible loss of power output.

3.4.2 Alignment

Most industrial lasers used in materials processing emit beams which are invisible to the human eye. Some laser systems, for example CO_2 laser medical systems, are fitted with a visible (HeNe) laser pointing beam.

Until recently, the components which allowed the two beams to be projected along the same path, called beam combiners, generally introduced lateral and angular deviations in the CO_2 beam, and so needed to be left in place after preliminary alignment.

A new, removable, version of this equipment is now available, designed to eliminate lateral and angular deviations of the working laser. Such a unit is removable after alignment is completed, and may then be used to align other laser systems or simply stored until required.

Finally, again causing no lateral or angular deviation to the working laser, a unit is available which directs a very small proportion of the working laser beam on to a heat-sensitive plastic film. This device renders the beam 'visible' at the screen, which is overlaid with a graticule to allow centration. In this case a subjective assessment of the working beam intensity profile can also be made.

3.4.3 Polarization – directing

As indicated in section 3.3.5, the use of circular polarization is a compromise, allowing the laser system to process metals equally but not optimally in all directions. V & S Scientific have built manually operated polarization rotators which, acting on a linearly polarized beam, allow the direction of the polarization to be changed. Although requiring further research, it is conceivable that a motorized version linked to a laser system CNC control could be used to direct the plane of polarization into the direction of the cut/weld, etc. and so give better results than in the case of circular polarization.

3.4.4 Focus-shaping

In conjunction with aberration-corrected lenses a variety of non-spherical components may be used to alter the shape of the focused spot.

For example, an axicon (conical surface) can be used to give a ring focus, perhaps of larger diameter for cutting circular holes in plastics, etc. or perhaps simply creating a small modification to the intensity profile at the focus to drill a hole of a specific shape.

3.4.5 Back-spatter reduction

Figure 3.11 shows a potential method of reducing the effect of back-spatter from the workpiece. The laser beam is focused, by either a lens or an off-axis parabolic mirror, as usual. A second, rotatable and replaceable auxiliary mirror is used to provide the final reflection of the beam on to the workpiece. Back-spattered material will degrade the auxiliary mirror surface, which may then be rotated to provide another clean section of

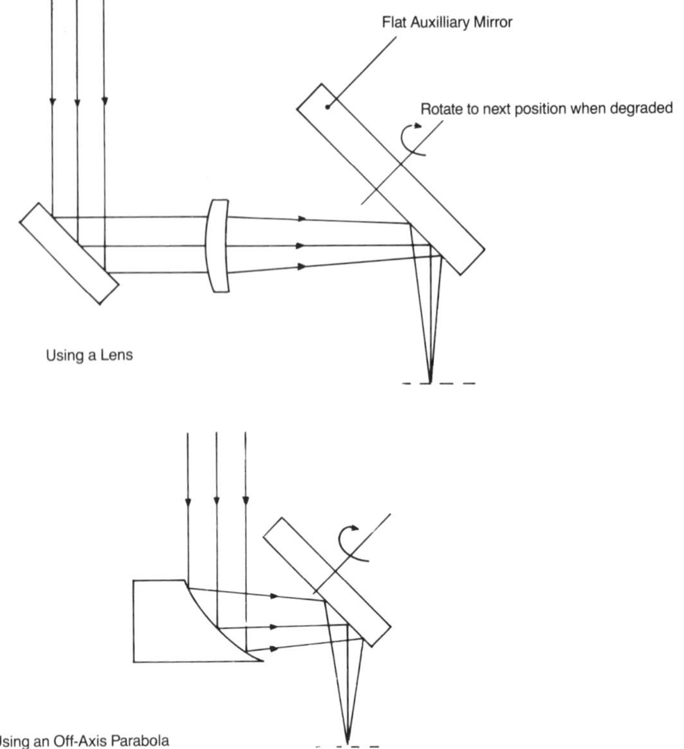

Fig. 3.11 Cost-reduction technique for high-back-spatter system.

its surface. In this way, it may be possible to use the auxiliary mirror in perhaps four or five positions before replacement or refurbishing is necessary.

3.5 SUMMARY

Until the late 1980s little thought had been put into the optical possibilities of laser beam delivery systems. Of course, certain institutions and universities were exceptions, but in general industry the trend was 'keep it simple'. This meant the use of some flat beam-directing mirrors where necessary, plus the use of a single element lens (often plano-convex) to focus the beam. There is a variety of applications when this is indeed sufficient, but many more where very large gains in efficiency can be achieved by the use of better optical design.

Much work remains to be done, but with improvements in optics such as the elimination of aberrations, and beam control using beam expanders, in addition perhaps to mode forcing and control of polarization, we are looking now at the possibility of building laser systems which achieve near maximum theoretical optical efficiency.

APPENDIX 3.A

We can show that the beam diameter at a beam waist multiplied by the far-field divergence is a constant factor.

NA = lens numerical aperture
D = beam diameter (mm)
F = lens focal length ('perfect' lens)
V_r = far field divergence in radians
V_m = far field divergence in milliradians = $1000 \times V_r$

The limit of resolution, Z = spot size, for a TEM_{00} beam is given by:

$$Z = 2\lambda/\pi \mathrm{NA} \quad \text{(A3.1)}$$

By definition $$Z = F.V_r \quad \text{(A3.2)}$$

The lens numerical aperture $\mathrm{NA} = D/2F$ (A3.3)

Substitute for NA, using equation (A3.3) in equation (A3.1): $Z = 4\lambda F/\pi D$

Substitute for Z, using equation (A3.2): $F.V_r = 4\lambda F/\pi D$

Cancel F, multiply through by D: $D.V_r = 4\lambda/\pi$ (A3.4)

But $D.V_r$ = beam diameter × divergence.

Convert to V in milliradians; for CO_2 lasers $\lambda = 0.0106$ mm

$$DV_m = \frac{4000\lambda}{\pi} = 13.4963 = 13.5 \text{ for a } CO_2 \text{ laser.}$$

For other types of laser the value DV_m can be found by using the appropriate value for wavelength.

APPENDIX 3.B

Example of a real system, with a cutting bed approx. 5 m long, 3 m wide; using a 1200 W CO_2 laser.
'Near point' of the cutting bed approx. 3 m from the laser.
'Far point' of the cutting bed approx. 11 m from the laser.

The CO_2 laser output conformed closely to the 'real beam' of Fig. 3.5, except we must consider that the full diameter of the beam was everywhere about 20% larger than the diameter measured at the $1/e^2$ points.

So, at 3 m from the laser the full beam was about 17.8 mm diameter. At 11 m from the laser the full beam was about 34.4 mm diameter.

The user intended to use a focusing lens of 38 mm diameter and 100 mm focal length (ZnSe meniscus type).

One point is immediately apparent: at the far end of the cutting bed, the beam is too large for the lens, once allowance is made for mechanical mounting.

Application of equations (3.4) and (3.5) indicates another problem – a change in performance from end-to-end of the cutting bed.

At the near point: power density = 28.14 kW/mm^2
At the far point: power density = 19.00 kW/mm^2
(due to lens aberrations from the increased beam diameter).

Worse, the depth of focus at the far point of the bed was only 27% of the depth of focus at the near point. $(17.8/34.4)^2 = 0.27 = 27\%$

A beam expander of × 1.5 magnification was installed at the laser head, where the 'full' beam diameter was 15.7 mm.

The output beam was now 23.5 mm diameter, with divergence reduced to 1.55 mrad.

The focus was set, on the expander, to produce a beam 'waist' at a distance of 15.2 m, (the beam waist being set to 23.5 mm diameter). This meant that, from the expander to this 'waist' – beyond the far point of the bed – the beam was everywhere approx. 23.5 mm diameter.

Application of equations (3.4) and (3.5) to the expanded beam now gives:

At the near point: Power density = 57.2 kW/mm^2
At the far point: Power density = 57.2 kW/mm^2

The depth of focus is everywhere $(17.8/23.5)^2 = 57\%$ of that of the original 'near point', (a reasonable compromise).

The beam is now able to pass through the system without 'clipping'.

APPENDIX 3.C

Windows and lenses for high gas pressures

The formula involved use of the value for 'rupture modulus' which is usually to be found in 'pounds per square inch' (lbf/in^2).
So, the formula is stated in imperial units.

T = edge thickness required (inches)
N = rupture modulus (lbf/in^2)
P = excess pressure (lbf/in^2)
r = half-diameter of unsupported window/lens (inches)

$$T = (1.1 \times P \times r^2/N)^{\frac{1}{2}} \tag{C3.1}$$

Equation (3.1) indicates the thickness at which the component will just withstand the pressure.
It is normal to use a safety factor. For example, for ZnSe, N = 6400 lbf/in^2.

Use N = 1600 for a safety factor of 4.
Use N = 3200 for a safety factor of 2.

Example

A ZnSe window is placed in a mount with 3 in clear aperture. It needs to withstand 10 bar pressure (145 lbf/in^2) with a safety factor of 4.

$$T = (1.1 \times 145 \times 1.5^2/1600)^{1/2} = 0.474 \text{ in} = 12 \text{ mm thickness}$$

APPENDIX 3.D

Suppose we have a laser of output power P, beam quality factor Q. Having eliminated aberrations, we can select a beam expander and an aberration-free focusing lens to give us any reasonable values of D beam diameter and F focal length we wish.

The spot size we will form will be given by equation (3.3b)

$$X = \frac{2Q\lambda}{\pi \text{NA}} = \frac{4Q\lambda F}{\pi D}$$

The area of this spot is: $A = \pi\left(\frac{x}{2}\right)^2 = \frac{4\lambda^2}{\pi}\left(\frac{F}{D}\right)^2 Q^2$

The average power density PD is $P/A = P\left(\frac{\pi}{4\lambda^2}\right)\left(\frac{1}{F/\text{no}}\right)^2\left(\frac{1}{Q}\right)^2$

Note that $F/D = F/\text{no}$, so:
(a) PD is proportional to $1/Q^2$
(b) PD is proportional to $1/F\text{no}^2$

We can 'trade-off' power density against depth of focus, for a given Q factor.

For example, if we double the F/no (insert a lens of twice the focal length, say) the depth of focus increases by $(F/\text{no})^2 = \times 4$. However, the spot size has doubled, so the average power density has decreased by 1/4.

4

Pulsed Nd:YAG lasers in manufacturing applications

T.M.W. Weedon

4.1 INTRODUCTION

Lasers for manufacturing applications have developed into reliable, proven tools that compete on an economic and productivity basis with the more traditional technologies. Nevertheless, their development, and that of the ancillary systems, is proceeding at a considerable pace. As the competing technologies are changing too, the optimum choice of equipment changes with time. On balance, the trend is toward the laser. The manufacturing processes considered here are joining (welding, soldering and brazing), cutting, drilling and heat treatment. The most important joining process is welding.

Two laser types dominate. The first, the Neodymium-doped Yttrium Aluminium Garnet laser (Nd:YAG), is the subject of this chapter; it is a solid-state laser. The other, the Carbon Dioxide Laser (CO_2), is a gas laser and is discussed in other chapters. Two other lasers will become important in the short-to-medium term. These are the excimer laser (a gas laser) and the laser diode (semi-conductor laser). The excimer laser is complementary to the others, for processing fine detail in polymer films. The laser diode, well known as the light source for CD players, will take over some soldering from Nd:YAG and CO_2 lasers. As their cost falls, they will begin to be used to excite Nd:YAG lasers, so increasing the service interval and reducing operating cost. Other laser types in development will not impact the processes discussed here before 1995. The Nd:YAG laser and the more mature CO_2 laser are complementary. Comparisons are made here to highlight their characteristics and applications, so that the Nd:YAG may be seen in context.

Nd:YAG lasers weld, in production, materials up to 3 mm thick, in pilot production materials up to 5 mm thick, and experimentally up to 15 mm thick. The welding rate is sensitive to the material thickness, varying from 10 m per minute or more for thin material, down to a few millimetres per

Laser Processing in Manufacturing. Edited by R.C. Crafer and P.J. Oakley.
Published in 1993 by Chapman & Hall, London. ISBN 0 412 41520 8

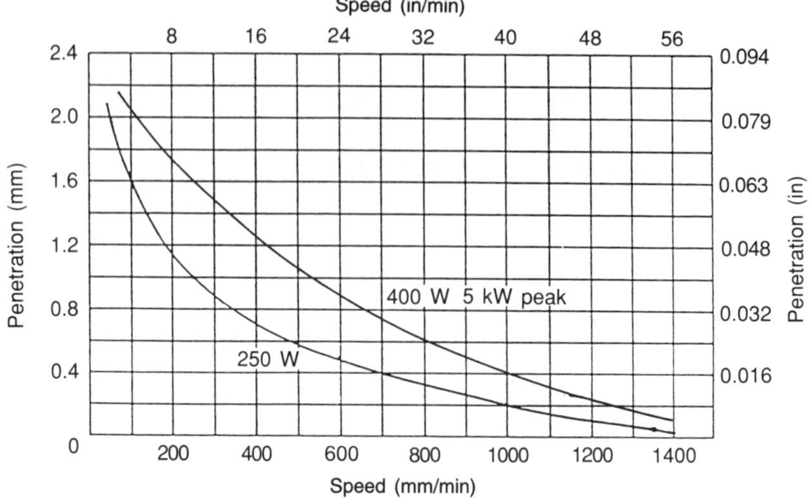

Fig. 4.1 Effect of penetration on welding feedrate for pulsed Nd:YAG lasers producing an hermetic seal in 304 stainless steel. Courtesy of Lumonics Ltd.

minute for thick sections. They cut, in production, up to 10 mm thick, and experimentally up to 25 mm or more. Again the feedrate varies from 10 m per minute down to a few millimetres per minute. Production drilling applications are for holes of diameter typically in the range 0.01 to 1.0 mm, through material up to 25 mm thick. Experimentally, holes are drilled through up to 100 mm, although this is very slow. Figures 4.1–4.3 indicate the effect of thickness on production rate. They should not be taken as an absolute indication of the productivity, as they are for good quality results and increased feedrates will result from lower quality requirements. Two key attributes of the Nd:YAG laser are its use with optical fibre beam delivery, and the versatility of modern programmable pulse-forming power supplies.

4.2 MARKET OVERVIEW

In 1988, the world market for pulsed Nd:YAG lasers of mean output power 150 watts and above was about 600 units while that for CO_2 lasers of 500 watts and above was around 1200 units (Mayer, 1989). Of these 1800 units, about 570 were installed in Europe, including 70 in the UK. The remainder was split roughly equally between the United States and Japan. By process, they were distributed as follows:

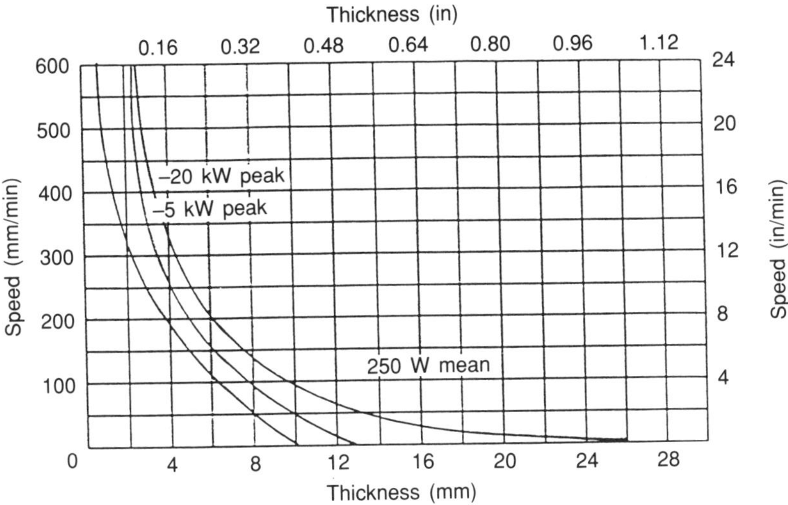

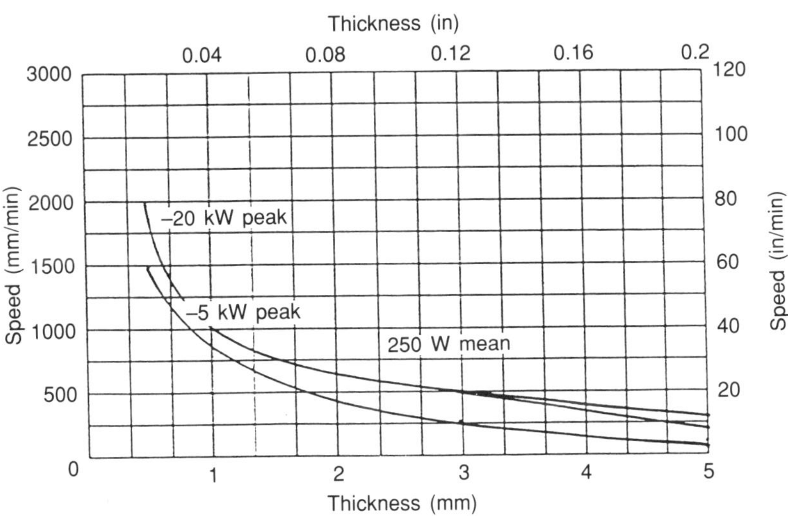

Fig. 4.2 Effect of thickness on cutting feedrate for pulsed Nd:YAG lasers and fixed optics in 304 stainless steel. Courtesy of Lumonics Ltd.

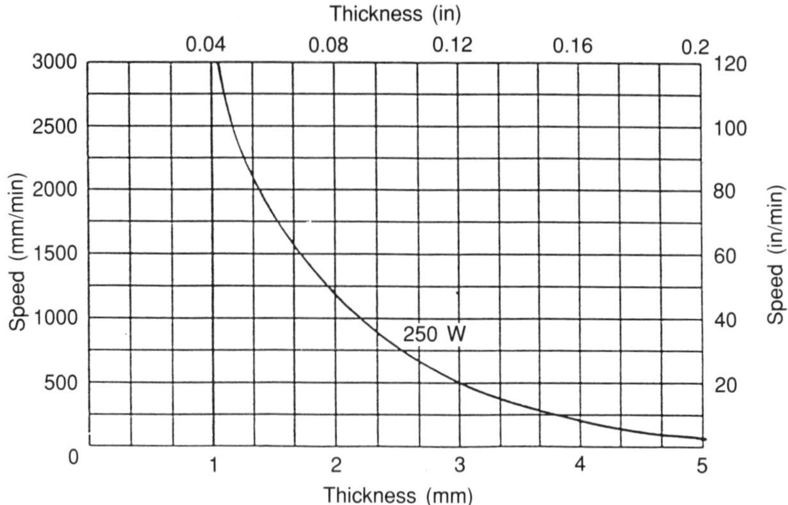

Fig. 4.3 Effect of thickness on cutting feedrate for pulsed Nd:YAG lasers and fibre optics in 304 stainless steel. Courtesy of Lumonics Ltd.

Process	CO_2 (%)	YAG (%)
Cutting	84	16
Drilling	0	12
Welding	15	72
Other	1	0
Total	100	100

The annual unit growth is around 10%, and the mix changes slowly. This sector of the laser industry is small, totalling about £360 million per year. Most CO_2 lasers are used for two-dimensional sheet cutting. Most Nd:YAG lasers are used for welding small electrical, electronics and fine mechanical parts. Making cooling holes in gas turbine components is the largest drilling application. Other markets are in the nuclear, space and automobile industries. As more powerful lasers become available, a welding market in the automobile industry will develop.

4.3 APPLICATIONS

4.3.1 Introduction

Lasers produce a near-parallel beam of 'light'. For Nd:YAG or CO_2, this power is in the infra-red. It is focused to a small spot on the surface of the work. Some or all of the light energy is absorbed by the material surface and its temperature rises. The temperature depends upon the material properties, the spot size and the peak power of the laser. It depends also on the laser type, because absorption is a function of wavelength. Broadly, the Nd:YAG laser is more suited to processing metals and some ceramics, and the CO_2 to organic materials (polymer, wood, paper and so on) and other ceramics.

4.3.2 Heat treatment

A few Nd:YAG lasers are used for heat treatment. This is where small, defined areas must be treated, for example to adjust the rate of a spring, improve local fatigue properties or produce a hard area suitable for grinding to a cutting edge. The material is heated to below its melting point, and then allowed to cool to yield the required result. The shape of the laser pulse, as a function of time, is customized to give the required thermal history to the treated area. The material may be remelted to get specific results. For example, cast iron can be remelted to dissolve the carbon precipitates and then allowed to cool rapidly to yield a hard homogeneous layer on the surface. Although this has been demonstrated, there are no known production installations. CO_2 or CW-Nd:YAG lasers may also be used for remelting.

4.3.3 Welding

Overview

Welding is a complex topic, because there are many approaches. All laser welding is fusion welding; a controlled region is heated above its melting point for long enough to allow material flow to produce the joint. The main attractions of laser welding are:

1. It is highly repeatable;
2. it is non-contact (although clamping may be needed);
3. there are no electrodes to wear or contaminate the weld;
4. it can be carried out in any chosen atmosphere;
5. the bulk temperature rise is small;
6. refractory metals may be welded since there is no limit to the weld pool temperature, other than that set by the melting and boiling points of the material;
7. dissimilar metals may be welded;
8. many geometries are suitable;
9. the laser beam is unaffected by electric or magnetic fields.

Laser welds are usually shielded locally with an inert gas, argon being the most common for Nd:YAG. Generally they are autogenous welds, but

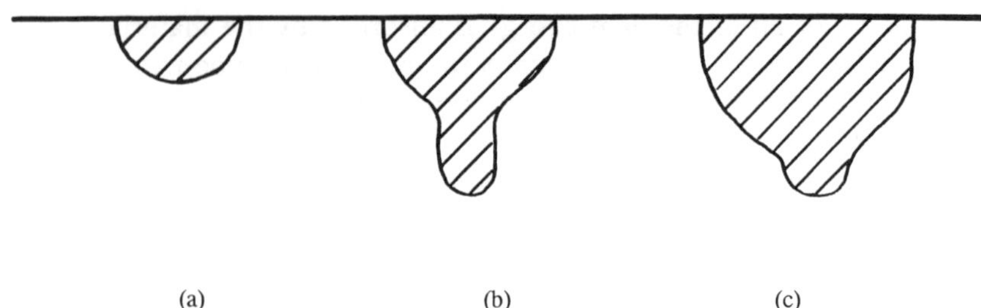

Fig. 4.4 Typical bead-on-plate welding results: transverse sections for (a) conduction limited weld at 400 W, (b) penetration weld at 400 W, (c) continuous molten pool weld at 800 W. Courtesy of Lumonics Ltd.

it is simple to provide a filler if required. This may include the use of a third component of suitable shape, wire feed or powder feed. Wire feed welding with the YAG laser is much simpler than with the CO_2, although there is no obvious explanation for this.

Spot welding

If the surface temperature is below the boiling point, heat transport is predominantly by conduction, and a 'conduction limited weld' is produced. The laser pulse duration must allow melting to the required depth. The result is a more-or-less hemispherical weld. If the power is slightly above that for conduction limited welding, boiling starts in the weld pool. A cavity forms in the centre, which allows the laser beam to interact with material below the surface. After the pulse, material flows back into the cavity and a 'penetration weld' results. Although some metal has boiled, little vapour is lost from the surface, normally much less than 1% of the weld volume. Hence the lost material has negligible effect on the weld, but may contaminate the surrounding atmosphere, and the parts of the machine close to the focus. Fume extraction may be needed.

Seam welding

A seam weld is produced by a sequence of overlapping spot welds. At low speed, the material cools between the pulses, and each spot is similar to a single spot weld, Fig. 4.4(a). As the average power and the feedrate are increased, the energy required for each spot falls because heat is retained from the previous spot. Eventually the weld does not solidify between pulses, Fig. 4.4(b). Then, if a penetration weld is being made, a 'keyhole' forms through the work. Such keyhole welding is normal for CO_2 (Banas, 1986).

Alternatively, if a conduction limited weld is made at high power, a 'continuous molten pool' (CMP) weld is produced, Fig. 4.4(c).

Penetration welds are energy efficient and lead to low bulk heating. They are narrow and deep, so they cause little distortion, but may be unsuitable for lap welds in thin materials. An exception is the welding of contaminated, plated or coated sheets, as the vapour from the contaminant may escape through the keyhole between pulses.

CMP welds are similar to Tungsten Inert Gas (TIG) welds, and appeal to the aircraft industry because they resemble welds for which a vast amount of knowledge, experience and testing exist. They allow the escape of volatile materials that might otherwise cause weld porosity. Their disadvantage is that they are energy inefficient and lead to increased bulk temperature and reduced throughput. They are broad compared with penetration welds, and can lead to distortion. The advantage of CMP welding, over TIG welding, is that the beam is not influenced by electric or magnetic fields, or by adjacent conducting features.

The selection of laser pulses for welding is important. The material must be held at the right temperature for the time to form the weld. For penetration welding, the thickness to be welded determines the peak power. The welding speed determines the repetition rate, and so the average power. The welding speed may be chosen according to various criteria; for example:

1. As fast as possible. When there is much welding to be done the number of lasers required is inversely proportional to feedrate;
2. as fast as possible while observing limits on the part temperature;
3. as fast as necessary for the production rate;
4. as fast as the work handling can cope with.

In the first case, if there are no material problems (cracking, surface contamination, etc.) then a CW laser of the appropriate peak power may be appropriate. Where bulk temperature is an issue, the pulse repetition rate is selected to yield an acceptable compromise between temperature and throughput. It is common, in the electronics and fine mechanics industries, to find welds that need high peak power, up to 5 kW say, but at a low throughput. The laser that meets these needs may have only a few tens or hundreds of watts average power. Manipulation problems, sharp corners for instance, may mean that parts of the joint have to be welded slowly, at reduced pulse rate. The use of a versatile laser that can cope with this may result in considerable savings in work-handling cost. Metallurgical problems may require detailed shaping of the laser pulse. For example, if the material is sensitive to cracking when quenched quickly, it may help if the pulse falls slowly. If it is susceptible to thermal shock, it may help if the pulse rises slowly. If the surface has a coating or is dirty, a low-level prepulse may evaporate the contamination before the welding pulse. For grain size control, quite complex pulse shapes are used.

4.3.4 Drilling

When the peak power far exceeds that required to boil the material, drilling results. Liquid and vapour are expelled from the focus region by the expanding vapour, revealing a new surface to be drilled further. The laser must reach this high peak power in short pulses, so that little heat is conducted sideways from the hole into the parent material. Although simple when expressed in these terms, laser drilling is the most complex of the manufacturing processes. The key quality issues are geometric tolerance and, for high-stress applications, metallurgical factors that influence fatigue life. These expand into:

- diameter
- roundness
- cylindricity or taper
- drilled surface roughness
- recast thickness
- recast composition and structure
- delamination
- microcracks

This is a different list from that for mechanically drilled holes. The manufacturing engineer must become familiar with parameters not used before. This tends to limit laser drilling to cases where the user has the relevant expertise, where no viable alternative process exists, or where there are so many holes to be drilled that all options have to be studied. The advantages that attract users to overcome these difficulties are:

- speed
- ability to process very hard materials
- non-contact process that does not have to cope with tool wear
- opportunity to drill at a low angle to the surface

The closest competing process is electro discharge machining (EDM), which may have a speed advantage when arrays of holes with parallel axes have to be drilled.

There are several approaches to laser drilling. Each has particular advantages. As there is considerable confusion over terms, careful definitions are needed. In 'single-shot drilling' a single laser pulse drills a hole whose diameter is determined by the focus diameter and the laser pulse parameters. It may be used with great ease on moving workpieces, so called 'drilling on-the-fly'. Drilling on-the-fly eliminates the repositioning time required if a part is brought to rest for each hole. An example is the de-icing panel used on the leading edge of some aircraft wings (Perun and Baker, 1988). Holes are drilled through titanium sheet at more than 50 per second.

Another on-the-fly application is the balancing of rotors of, for example, dental turbines or high-speed electric motors. The rotor is run on a balance measuring machine with feedback control to the laser. Material is removed from the appropriate places. This is particularly valuable for flexible rotors, for example in ultra centrifuges, where mass must be removed from many planes. Drilling on-the-fly causes some elongation of the holes. This depends on the pulse duration and the part velocity. It is exploited in balancing applications to yield a long shallow crater. 'Percussion drilling' is similar to single-shot drilling except that multiple pulses are used. Normally the part is stationary during drilling, but sometimes the hole is produced by successive passes over the part, drilling on-the-fly. Percussion drilling and single shot drilling are known collectively as 'direct drilling'.

It is easier to drill through-holes than blind holes. All the more so if a gas nozzle is used. A coaxial gas nozzle is employed for two reasons. First, it reduces the recast layer in the hole bore, by increasing the shear force to the liquid film. It also reduces the impingement of debris on the focus lens.

Often, oxygen is chosen as the assistance gas, because it is more effective than air in reducing the recast layer. Its effect seems greater for materials whose oxide acts as a flux, reducing the viscosity of the liquid metal. Oxidation of the metal vapour and liquid contributes energy to the process, a factor important in CO_2 laser cutting. It is probably not significant in Nd:YAG laser drilling. The gas nozzle has the disadvantage of increasing the drilling time up to the point where the hole breaks through the work. Its benefit is the improved quality of the hole after it breaks through. When high precision is required, trepanning is common. A central hole is direct drilled, and the required hole is cut from the workpiece by moving the focal spot. Typically the focus will be around 0.18 mm diameter, leading to a kerf of 0.2 mm width. This is suitable for hole diameters from 0.25 mm upwards and material thickness of 0.25–20 mm.

Trepanning, Fig. 4.5, almost always employs a coaxial gas nozzle, otherwise there is a high probability of liquid debris rewelding within the kerf. Trepanning is used to produce circular and non-circular holes, a requirement in some fluid-flow applications. It also allows simple hole diameter control, often in response to in-process measurement. An example is the calibration of idle air-flow apertures in fuel metering equipment. The laser that is suitable for trepanning has a great depth of focus, since it is required to cut deep with a narrow kerf. This is a disadvantage when producing holes in parts with a small internal cavity, such as is common with gas-turbine aerofoil components. The beam may pass through the front wall and damage the back wall, Fig. 4.6. This is unacceptable where a high working stress and long fatigue life are required. Unfortunately, this is just the situation where the high quality hole produced by trepanning is required. An insert may be fitted into the cavity to absorb or diffuse the laser beam

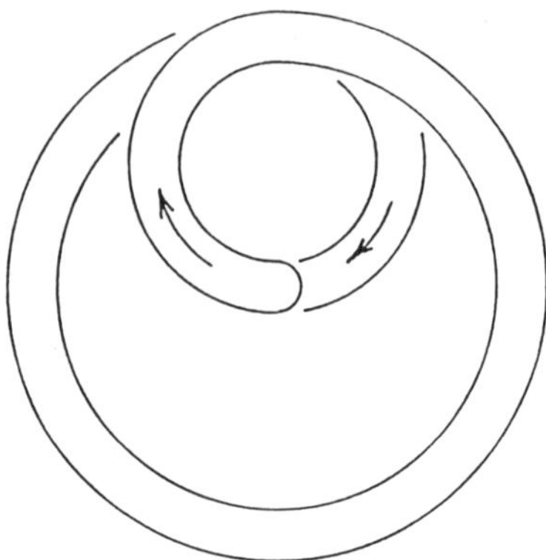

Fig. 4.5 The trepanning trajectory for a circular hole showing the centre start from a direct drilled hole. Courtesy of Lumonics Ltd.

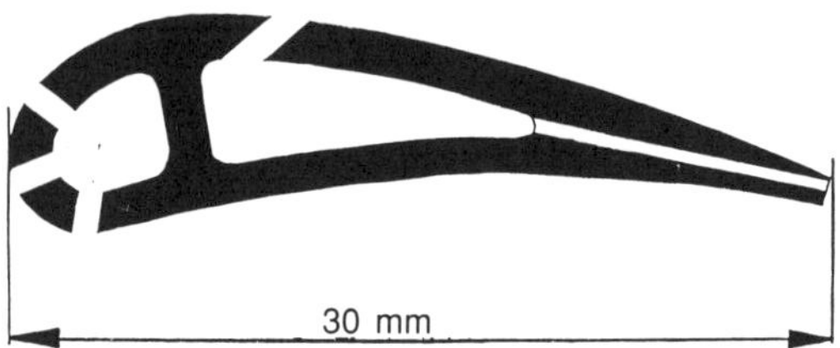

Fig. 4.6 Schematic section of a gas turbine aerofoil component showing the proximity of the back wall to the exit face of the front wall. Courtesy of Lumonics Ltd.

to prevent back-wall damage. This must permit the gas assistance to function, by allowing flow out of the cavity, and must be removable after drilling.

4.3.5 Cutting

Cutting is effectively drilling overlapping holes, usually with a gas nozzle. It is simple to optimize, reliable and repeatable. Cutting metal sheets in two dimensions is generally more competitive by CO_2 lasers, for which the energy contribution from oxidation of the metal is higher than for the Nd:YAG laser. However, the freedom of beam manipulation offered by fibre optics makes the Nd:YAG laser attractive for three-dimensional applications. The high peak power of the Nd:YAG laser is an advantage in cutting thick metal, and results in a reduced heat affected zone in thin parts. For armour applications subsequent heat treatment or edge grinding may be eliminated.

4.3.6 Other processes

Soldering is like conduction-limited welding, except that lower pulse power and longer duration are required. The solder may be provided as a preform, plated on to the components, or as a paste; the flux may be a paste or a gas. The absorption of solder or tinned components is excellent at the Nd:YAG laser wavelength. Brazing is similar to soldering.

A novel process, originated in a European collaborative project (BRITE project, 1987) has been developed by Maho Machine Tools and others. Maho have called this 'Laser Caving'. The equipment is similar to that used for gas-assisted percussion drilling, and the focus is rastered over the work to 'mill' the material to depth. Various workers have used Nd:YAG or CO_2 lasers, but it is not clear which laser will prove appropriate for which materials.

4.4 ECONOMIC ASPECTS

The capital cost of Nd:YAG lasers is more sensitive to mean power than to peak power, and is typically in the region £150 to £200 per watt. To this has to be added the provision of space (a 400 W laser has a footprint of about 1.2 m^2 including its power supply), of electrical and cooling supplies, and possibly fume extraction. A machine will be required to move the focus relative to the workpiece. The capital cost of the Nd:YAG is around twice that of the comparable CO_2 laser. The beam delivery for the Nd:YAG is however much cheaper than that for the CO_2 laser, especially for five-axis applications. In consequence the total system cost may be lower than for CO_2. Running costs comprise consumable and maintenance costs. The consumables are the three main services, electricity, cooling and gas (argon for welding, air or oxygen for cutting and drilling), plus flashlamps. Of these, the cost of gas dominates in cutting and the cost of flashlamps dominates in welding. Maintenance consists of changing the flashlamps when necessary. Every six months it is desirable to replace the filters in the laser cooling system, and to inspect the laser head optics. The glass slide that protects the focus lens will need to be replaced at intervals between half a shift and one month, depending upon the process and the beam delivery system.

4.5 SAFETY AND ENVIRONMENTAL ASPECTS

The risks to be considered are: optical; electrical; fume; fire/explosion.

The laser beam is powerful enough to cause burn injuries to the operator's skin or clothing. It is not an ionizing radiation, so it does not cause cancer. The Nd:YAG laser wavelength is focused by the lens of the eye, so scattered laser light can damage the retina to the extent of causing blindness. These hazards are controlled by administrative and design standards (IEC 825). The observance of this standard is mandatory in the United Kingdom, under the Health and Safety at Work Acts, and under equivalent legislation in other countries.

Electrical hazards similarly are controlled by standards and legislation in most countries. The relevant international standard is (IEC 820).

Laser materials processing may produce toxic, flammable or explosive fumes or dust. It is therefore essential that the user identify the composition of the workpiece, and understands the hazards associated with such material. Some particular cases are:

1. Metal dust can be explosive, particularly in an oxygen atmosphere, and can accumulate in ducts and filters. Here it may be ignited by a stray spark from the process or from static electricity generated in the duct. Most hazardous are the reactive metals, for example magnesium, zirconium, titanium and to a lesser extent aluminium.
2. Many workpieces are flammable, particularly in an oxygen atmosphere. Magnesium, zirconium and titanium should be cut only with inert gas assistance.
3. Beryllium oxide is highly toxic, and is likely to be produced when cutting or welding alloys containing beryllium, even in an inert atmosphere. Expert advice should be sought.
4. Polymers and polymer-matrix composites can produce toxic fumes. It should always be assumed that these materials demand excellent fume extraction.

Dust and fumes are undesirable from a health and safety viewpoint, so the exposure of personnel should be reduced to the absolute minimum.

As compressed gases are used in most laser processes, the normal precautions are required.

Laser manufacturers maintain an awareness of the safety aspects of lasers and their applications, and are happy to help their customers with these matters. Guidance is available from regulatory bodies (the factory inspector in the UK), standards organizations, and industry associations. TWI Laser Centre in the UK, and the Laser Institute in America both offer help and training. The absolute responsibility on the user to ensure the safety of his process is not a major burden in the application of laser processing.

4.6 CASE STUDIES

Some examples of the use of Nd:YAG lasers in manufacture have been mentioned. The following thumb-nail sketches outline applications that have been publicized by users. They give interesting messages about lasers in manufacture.

4.6.1 Thorn Lighting H4 headlamp

The application is spot welding 0.25 mm thick molybdenum sheet components, and has been in operation for more than fifteen years. Molybdenum has a high softening temperature. This makes resistance welding difficult. The existing technology was brazing with platinum, using a resistance welder as the energy source. The operational problem was the short and unpredictable life of the welder electrodes. This varied up to a few hundred parts. At the production rate of a few thousand parts per hour, the maintenance time made automatic machines uneconomic. Using a Nd:YAG laser not only eliminated this maintenance, but dramatically reduced the standard deviation in weld strength, so reducing the scrap rate significantly. Consequently it was possible to automate the lamp-making machine and Thorn and their customers benefited from improved quality and production cost.

This case is interesting, as it exemplifies a management approach that has yielded great benefits to the user. Having proven the technical feasibility in laboratory trials, Thorn installed a manually loaded laser welding station in their pilot plant, and used this to optimize the process. They then built an automatic machine, also operated in the pilot plant, and added further operations incrementally. When the system was fully operational in this 'sheltered' environment, it was introduced to full production (Weston, 1985; Howe and Morris, 1988). This approach has been adopted for first-of-a-kind machines by many companies in many countries. It seems to be the route that yields the greatest benefits, and is much more successful than attempts to reduce the number of process development steps, with a consequently less optimal solution.

4.6.2 Heart pacemaker package sealing

Heart pacemakers operate in a constant-temperature environment. To withstand the body moisture, and to be compatible with body tissue, they are encased in sheet titanium. This was previously sealed by electron-beam welding. The process took place in a vacuum chamber, and resulted in a high bulk temperature rise, which had two results. First, the electronic components had to be selected to withstand the temperature, and had to be protected by thermal insulation. Second, the high temperature caused the encapsulation to out-gas, causing the incidence of porous welds.

Medtronic switched first to CO_2 laser welding, then, in the mid-1970s, to Nd:YAG laser welding. This allowed them to eliminate the thermal insulation and to adopt low-power electronics technology that was more heat sensitive. A dramatic size reduction resulted from the reduced insulation and battery volume. Subsequently, other manufacturing steps have been changed to Nd:YAG laser processing, as described in papers by authors from Medtronic (Janssen, see Fig. 4.7) and others.

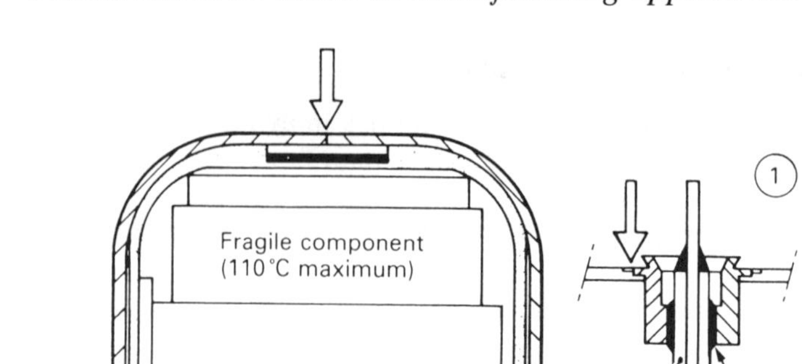

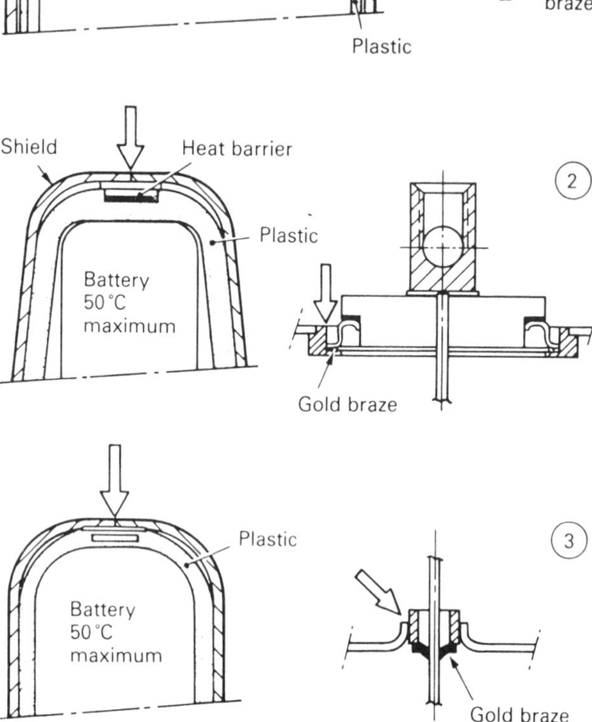

Simplification of mechanical design in three phases; large arrows indicate beam incidence direction

Fig. 4.7 The evolution of pacemaker design. Taken from Janssen (1984). Courtesy of TWI.

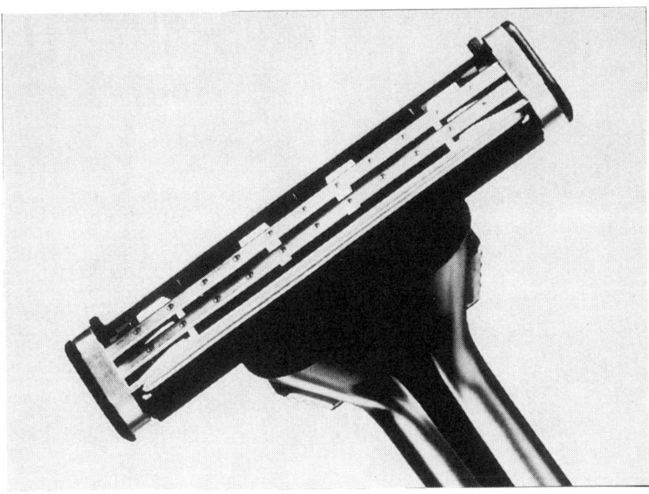

Fig. 4.8 The Gillette Sensor Razor showing the laser spot welds that attach the blades to their supports. Courtesy of Lumonics Ltd.

4.6.3 Gillette 'Sensor' razor

Considerable publicity has been given to the new razor that Gillette introduced to the United States and Europe during 1990. Apart from being a fascinating story of product design and development, this consumer product is an interesting example of the profitable use of the Nd:YAG laser for high-volume, high-reliability welding. Thirty lasers, with fibre-optic beam delivery systems, are integrated with sophisticated production and inspection systems to produce a total of three million welds per hour (Chesnaukas and Llewellyn, 1990), (Fig. 4.8).

The benefits of Nd:YAG laser welding were:

1. High speed and the ability to get a constantly acceptable weld at varying speed;
2. process cleanliness coupled with tolerance to imperfect component cleanliness;
3. ease of integration with the other process equipment, not least because of the use of fibre beam delivery;
4. compactness, less than 50% of the footprint of the alternate CO_2 equipment;
5. high reliability of the process and laser;
6. low operating cost because of low consumable cost;
7. on-line diagnostics, a feature of microprocessor-controlled transistorized Nd:YAG power supplies, allows pro-active in-process quality control.

Fig. 4.9 Experimental laser-fibre-robot cutting cell at Volvo Car Corporation. The multiple lasers, fibres and beam switching units are visible in the background. Courtesy of Volvo Car Corporation.

4.6.4 Car-body cutting

Over the period 1987–9, the Volvo Car Corporation developed a pilot-plant cutting cell for the production of prototype parts, and customization of standard vehicles, Fig. 4.9. The system comprised an articulated arm robot, fibre-optic beam delivery and assist gas nozzle with height sensor. This was integrated through a cell control computer to the central CAD system. The part geometry was processed by the CAD computer to produce the robot part programme, which was then reprocessed by the cell control computer to integrate the laser parameters from the process database. This system, the viability of the equipment and the approach, has been the subject of papers by Hanicke (1988) and Wallander (1989). Volvo have shown high-quality cutting in a range of car body materials, including finished painted, zinc-coated steel.

4.6.5 Other cases

The literature contains many examples, including the notable use of Nd:YAG lasers in the production of cooling holes in gas turbine combustion chambers and aerofoil components. Useful references are Corfe (1983); *Machinery* (1983); van Dijk *et al.* (1989); Weedon (1984).

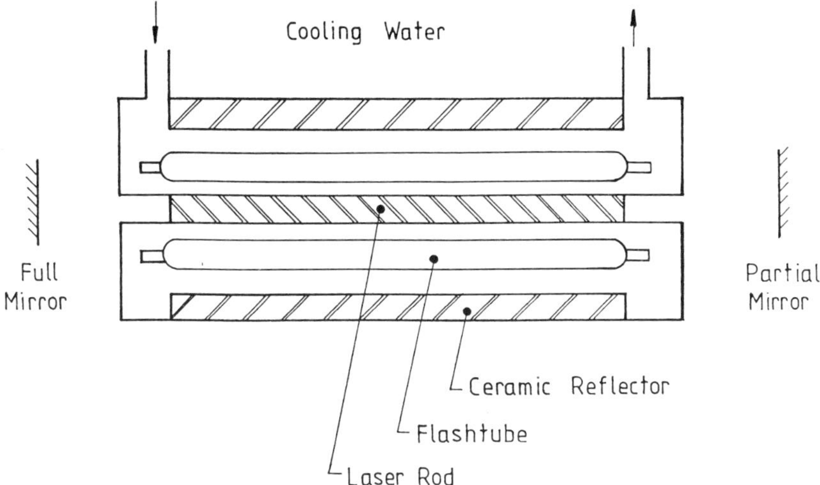

Fig. 4.10 Schematic of a simple solid-state laser oscillator. Courtesy of Lumonics Ltd.

4.7 THE PULSED Nd:YAG LASER

The heart of the laser is the rod of Neodymium-doped Yttrium Aluminium Garnet (Nd:YAG), which acts as a light amplifier at the specific wavelength 1064 nanometers. This wavelength is in the near infra-red, and is transmitted by the materials commonly used for visible light optics. It is also transmitted by the lens of the eye, even though it is not visible.

YAG is a diamond simulation material and is the hard, optically excellent host for the Neodymium that is the laser active medium. It is strong and has good thermal conductivity. These important properties for laser rods effectively set a limit to the average power that can be extracted from the rod. The laser rod is excited, or 'pumped' by light that is transmitted into the rod through its cylindrical surface. This light is provided in material-processing lasers by flashlamps similar in design to large photographic flash lamps. In future, some such lasers will be pumped by laser diodes, but this development will be paced by the availability and price of suitable diodes. It will probably be after 1995 before many lasers in manufacture use diode lasers.

Figure 4.10 is a schematic of the laser oscillator that results when an excited laser rod is provided with positive feedback. In this sketch, the feedback is by two parallel mirrors that reflect the laser light repeatedly through the laser rod, allowing it to be amplified on each pass. The laser beam escapes from the resonator through the partially reflecting mirror, or 'output coupler', to be available for useful work. The sketch also represents the reflector and housing that enclose the laser rod and lamps. These maximize the efficiency of the excitation of the rod, and provide the means

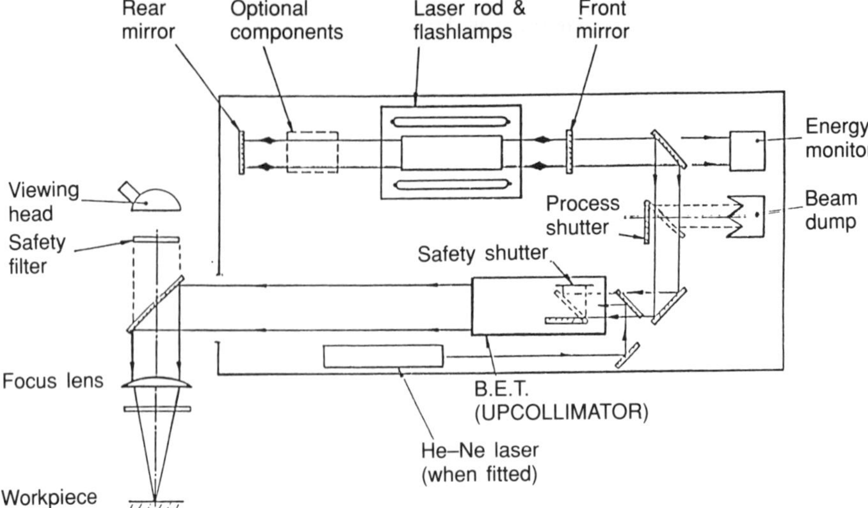

Fig. 4.11 Schematic of an Nd:YAG laser head (JK701) shown with a simple fixed optic beam delivery. Courtesy of Lumonics Ltd.

to direct a cooling water flow over the components. This is necessary because the efficiency is poor at best, meaning that there is much waste heat to be removed. The laser rod is typically 150 mm long, and 5–10 mm in diameter. The laser oscillator is housed, with other essential optical, control and safety features in the laser head. This is shown schematically in Fig. 4.11, and in a practical form in Fig. 4.12. The zoom beam expanding telescope allows the operator to fine tune the focus spot size to meet the needs of the application.

4.8 BEAM DELIVERY SYSTEMS

Until the mid-1980s, most high-power Nd:YAG laser applications relied on 'conventional' or 'fixed' optics to carry the beam from the laser to the work, shown in a simple form in Figs. 4.11 and 4.12. In the early 1980s, Nd:YAG lasers began to be used with fibre-optic beam delivery. This approach has grown so that now most material-processing lasers are equipped with fibres.

Fixed optics rely on the use of mirrors and lenses to manipulate the laser beam, and to focus it on to the work. Such systems rely on a rigid common support for the laser, the beam delivery components, and the work. This is straightforward for modest-size workpieces that may be processed close to the laser. The relative motion of the focus and the work is by moving the focus, the work, or both. Generally there are few problems

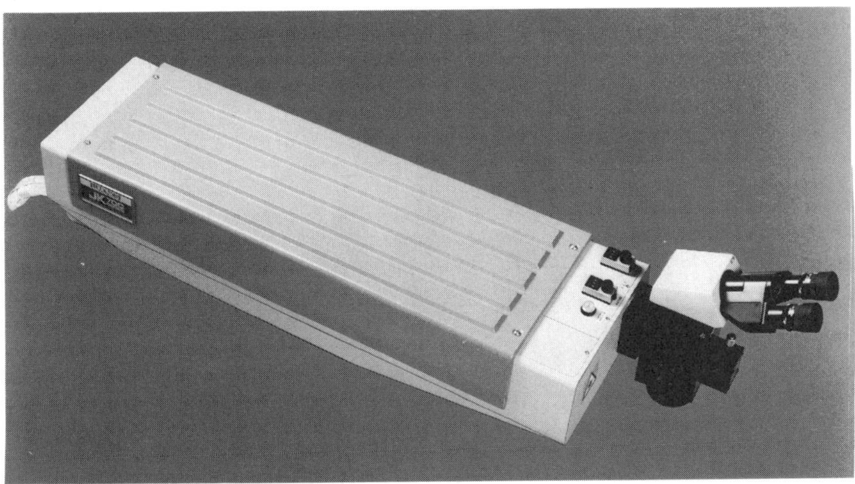

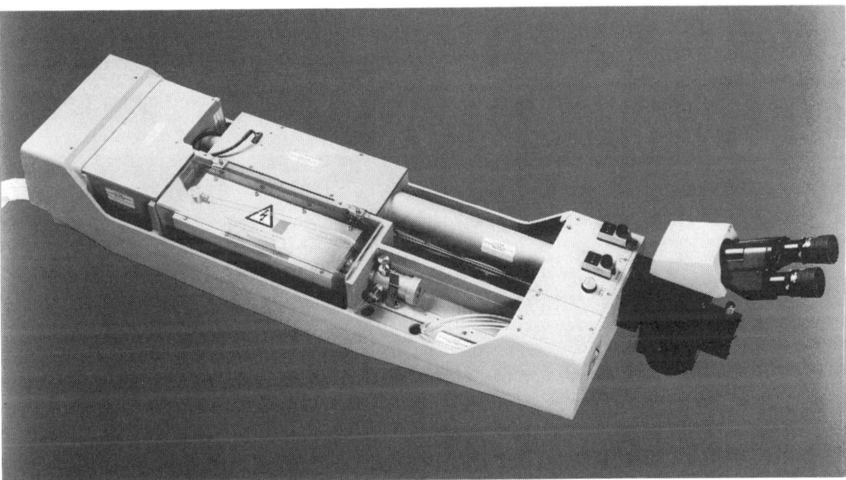

Fig. 4.12 A 400 W pulsed Nd:YAG laser (JK701) shown with and without its cover. The controls are for the zoom beam expanding telescope. The laser rod and flashlamps are housed in the 'pumping chamber' on the near side of the photograph. Courtesy of Lumonics Ltd.

in translating the beam in three axes, but it needs special arrangements to handle the beam in five axes (Fig. 4.13). It is often more cost effective to rotate the part (Fig. 4.14), although this may cause the machine to be larger. Detailed analysis is required in each case to establish the optimum solution. Fibre-optic beam delivery is a much more flexible solution for those cases where it is appropriate. These are welding, cutting where a

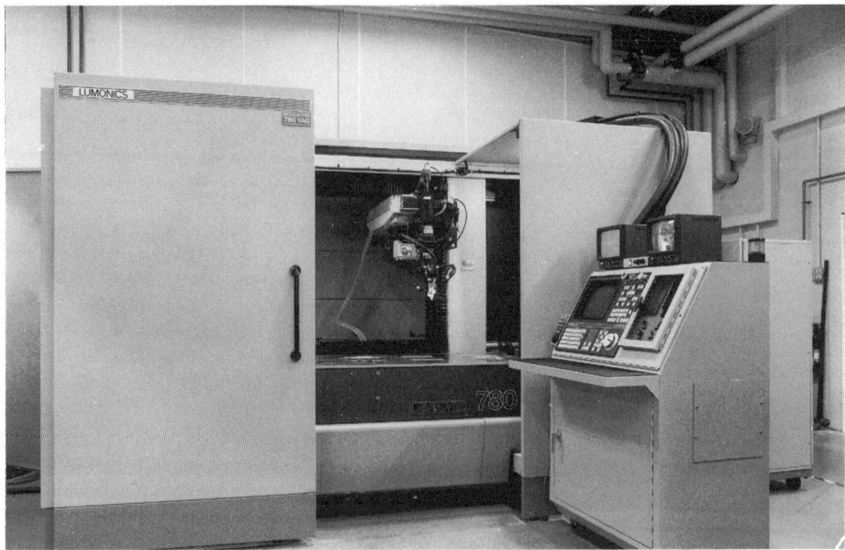

Fig. 4.13 Laserdyne 780 Beam Director with JK704 showing the class 1 safety enclosure, the laser head, and the twin optical rotary axes using fixed optics. Courtesy of Lumonics Ltd.

Fig. 4.14 Seven-axis CNC laser drilling system for gas-turbine components, showing the programmable hole shape trepanning unit, and rotary/tilt table that manipulates the part in two axes of rotation. Courtesy of Lumonics Ltd.

very narrow kerf is not required, and drilling thin sections. There is some loss of laser energy coupling the laser beam into and out of the fibre, but negligible loss along its length. Thus it is feasible, and quite common, to operate the laser at some considerable distance from the work-station. This has great advantages where there is a hazard associated with the work, for example an explosion risk. The laser and the operator may be placed behind a blast wall at a distance from the work.

It is advantageous to be able to do clean-room welding with the laser in a remote plant room where space is less costly. Fibres up to 50 m length are used routinely. Fibre-beam delivery offers greater freedom to share the laser between several working points, either by 'energy sharing' or by time multiplexing the output. It also offers the opportunity to combine the outputs of several lasers into one focus (Lumonics patent applied for). As the fibre is mechanically flexible, and movement does not affect the work that is done with the laser, there may be relative motion between the laser and the working point. This may be quite limited, for example vibration or flexure of a factory floor, or it may be gross as when the fibre output is manipulated with a robot. Figure 4.15 illustrates some of the beam delivery modules that are employed with Nd:YAG lasers.

4.9 TRENDS

As production use increases, laser manufacturers will be able to reduce the capital cost of systems. There has been an enormous improvement in effectiveness/cost over the last decade, and one can expect this trend to continue. Higher-power lasers are needed for new applications, and to increase the speed of existing operations. The new tasks demand even higher reliability than the 98–99% uptime typical of Nd:YAG lasers of 400 W mean power. When this is achieved, there will be a spin-off into the medium-power equipment. The coming of diode pumping of Nd:YAG lasers has already been mentioned.

Fibre-optic beam delivery is less developed than the laser technology it serves. There will be further development to yield more cost-effective beam delivery, with the capability to deliver smaller spots. This will result ultimately in the almost universal adoption of fibre-beam delivery. The expansion of laser use in manufacture is limited by the unfamiliarity of design and manufacturing engineers with the characteristics and capabilities of laser processing. This is recognized by the governments of many developed countries, and by more forward-looking large companies. We will see a period of technology transfer from the current practitioners to potential new users. Laser manufacturers have a role in presenting the process characteristics of their products in the terms with which potential users are familiar. Whilst this appears obvious, it is not simple.

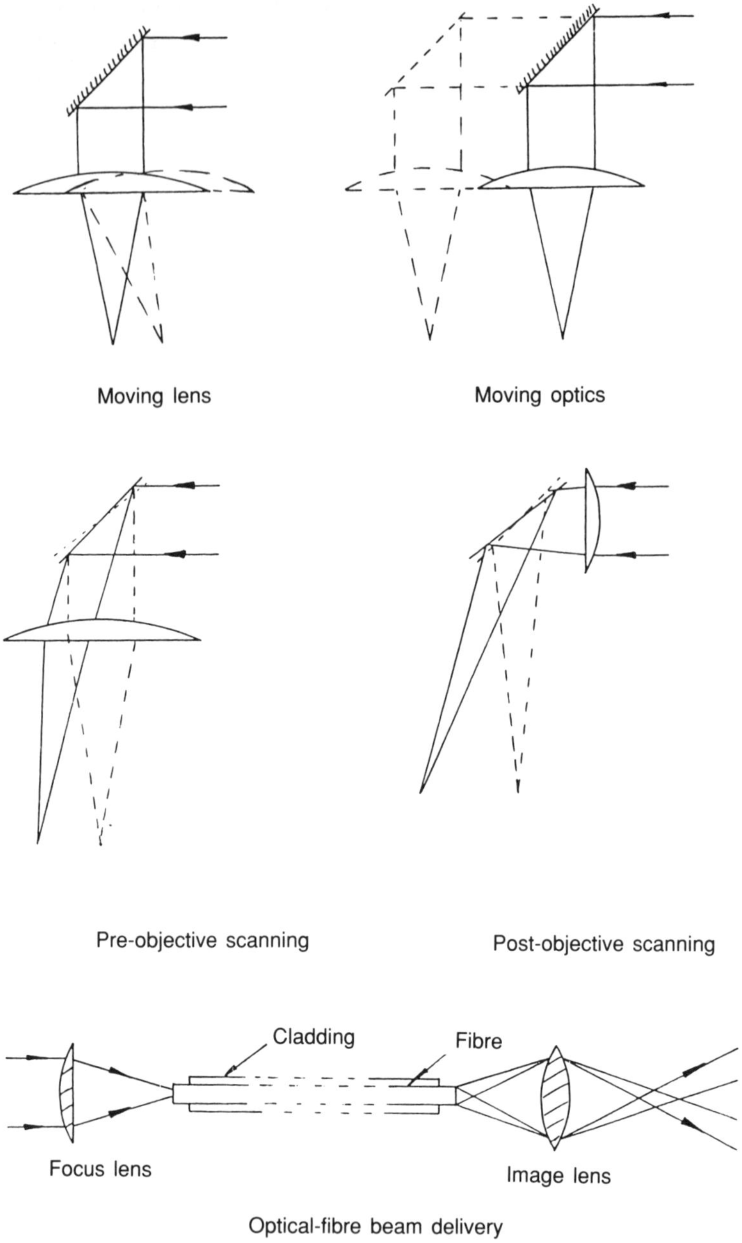

Moving lens

Moving optics

Pre-objective scanning

Post-objective scanning

Cladding

Fibre

Focus lens

Image lens

Optical-fibre beam delivery

Fig. 4.15 Various methods of moving the focus spot relative to the work. Some of these are inappropriate to the use of a gas assistance nozzle. Courtesy of Lumonics Ltd.

REFERENCES

Banas, C. (1986) High-power laser welding, in *The Industrial Laser Handbook, 1986*, Penn Well Books, Tulsa.

BRITE project 1206 (1987) Presented annually at BRITE Technological Days, Brussels.

Chesnaukas, R. and Llewellyn, S. (1990) Laser welding makes it a close shave. *Industrial Laser Review*, May, 4–5.

Corfe, A.G. (1983) Why a laser is better than EDM for drilling. *Production Engineer*, December.

Van Dijk, M.H.H., de Vlieger, G. and Brouwer, J.E. (1989) Laser precision hole drilling in aero-engine components. *Proc. IFS Conference*, Birmingham, May, LIM-6.

Hanicke, L. (1988) YAG-fibre-robot system for laser cutting in car body steel. *Proc. IFS Conference*, Stuttgart, September, LIM-5.

Howe, S. and Morris, R. (1988) Lasers in lamp making: ten years' experience in Thorn Lighting, Ltd. *Proc. IFS Conference*, Brighton, LIM-4.

IEC 825 (1984) *Radiation Safety of Laser Products*. International Electro-Technical Commission, Geneva, Switzerland.

IEC 820 (1986) *Electrical Safety of Laser Products*. International Electro-Technical Commission, Geneva, Switzerland.

Janssen, G.W.G. (1984) Laser welding in the manufacture of heat pacemakers, in *Laser Welding, Cutting and Surface Treatment* (ed. R. Crafer), The Welding Institute, Cambridge.

Mayer, A. (1989) *Lasers 2000 – Update*, Baasel, October.

Perun, K.R. and Baker, K.J. (1988) New laser perforating process for titanium. *Proc. Sixth World Conf. on Titanium*, France.

Wallander, D. (1989) New laser applications at Volvo Car Corporation. *Proc. Second NOLAMP Conf.*, Lulea, Sweden, August.

Trepanning moves in on mini holes (1983) *Machinery*, November.

Weedon, T.M.W. (1984) The design of a CNC laser drilling machine for the production of high quality holes in aerospace components. *Lasers in Manufacturing*, April, Hertford, Conn.

Weston, R.F. (1985) Lasers in lamp technology. *Lighting Journal*, **13**.

GENERAL LITERATURE

The following are useful references, but please note that not every author recognizes that results may be highly specific to a particular laser. The performance or problems described may vary from one equipment to another.

The Industrial Laser Handbook (eds. D. Belforte and M. Levitt), published annually, PennWell Books, Tulsa, Oklahoma, USA.

Laser Focus World, published monthly. PennWell Books, Tulsa, Oklahoma, USA.

Industrial Laser Review, published monthly. PennWell Books, Tulsa, Oklahoma, USA.

Lasers and Optronics, published monthly. Gordons Publications, Morris Plains, New Jersey, USA.

Proceedings of The Lasers in Manufacturing Conference. The first conference was held in Brighton in 1983, and there have been six to date. IFS Publications, Bedford.

Proceedings of various specialist and general conferences organized by the SPIE.

Proceedings of ICALEO conferences organized by the Laser Institute of America.

5

Continuous wave and *Q*-switched Nd:YAG lasers

A.B. May

5.1 TECHNICAL BACKGROUND

5.1.1 Why are we interested in Nd:YAG lasers?

YAG lasers are solid-state lasers which means that unlike large CO_2 lasers, no gas circulation is required. The consumable is not gas but an arc lamp, almost as easily replaceable as a light bulb. YAG lasers are relatively compact, and though inefficient in terms of energy conversion (a few per cent at best), their *Q*-switchability permits enhancement of peak powers, enabling them to perform some very demanding tasks. Their adaptability is enhanced due to their efficient transmission by fibre optics. They have found many varied applications throughout the world, some of which will be described in the following pages, namely:

1. Resistor trimming – probably the biggest application for solid-state industrial lasers.
2. Laser marking or engraving – permanent, fast marking of a variety of materials.
3. Laser cutting and drilling.
4. Laser soldering – an elegant means of non-contact soldering arising from the increasing application of SMD (surface mount devices)-based circuitry.

5.1.2 What is a Nd:YAG laser?

Neodymium (Nd) is a rare earth element within whose energy level structure the laser transition takes place (see section 5.1.3 for details). 'YAG' is often used as an abreviation for Nd:YAG, and itself stands for Yttrium Aluminium Garnet. Pure YAG, $Y_3Al_5O_{12}$ is a colourless, optically isotropic crystal which possesses a cubic structure characteristic of garnets. In

Laser Processing in Manufacturing. Edited by R.C. Crafer and P.J. Oakley
Published in 1993 by Chapman & Hall, London. ISBN 0 412 41520 8

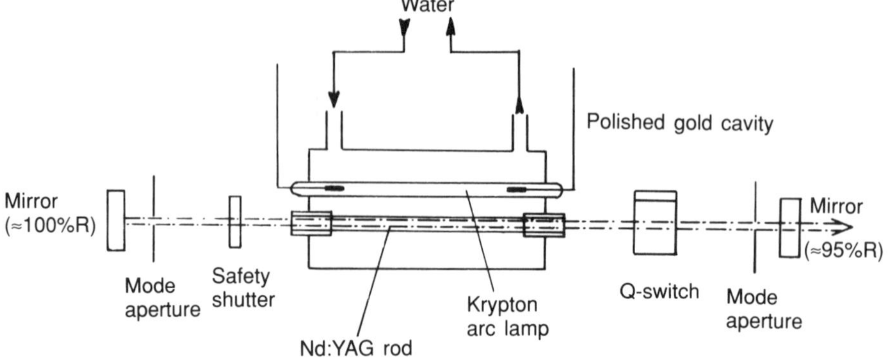

Fig. 5.1 Schematic of laser oscillator and pump chamber.

Nd:YAG, about 1% of Y^{3+} ions are substituted by Nd^{3+}. No wonder then, that 'YAG' is a popular term for 'Neodymium-doped Yttrium Aluminium Garnet'.

The YAG acts as a host lattice for the Nd^{3+} ions, and its physical, chemical and mechanical properties make it a most attractive choice. It is hard, has good optical transmission at wavelengths of interest and has high thermal conductivity. When doped with Nd, YAG rods look like a very light pink glass.

A typical Nd:YAG laser configuration is shown in Fig. 5.1. The most common arrangement is a pump chamber of elliptical cross-section with the rod at one focus and the pump lamp at the other. Double-ellipses are also used. The arc lamp is krypton-filled. Krypton is a noble gas whose emission spectrum closely matches the important absorption lines of Nd. This ensures efficient optical pumping and the cavity geometry ensures that the output from the lamp couples into the rod effectively. Cavities are generally either gold coated or ceramic. Both absorb little of the pump power, and some argue that the non-specular scattering of the ceramic ensures more diffuse illumination of the YAG rod, avoiding an axial 'hot-spot'. The cavity is cooled by recirculating de-ionized water.

The oscillator is defined as the part of the laser between the mirrors. The output coupler is the lower reflectance mirror, the transmission representing the laser output. Focusing optics for Nd:YAG lasers are usually made of anti-reflection coated glass, another attraction of a laser which operates at 1.06 μm. The *Q*-switch is discussed in section 5.1.4. One or more apertures are used to eliminate higher-order transverse modes from the laser. A mechanical shutter is normally incorporated for safety reasons.

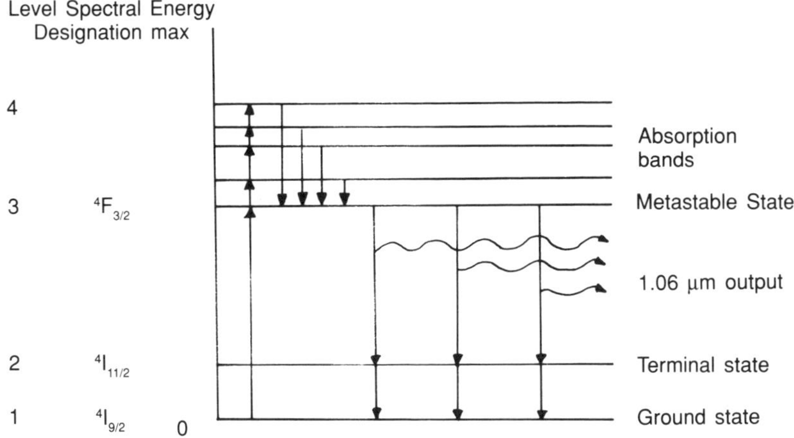

Fig. 5.2 Energy-level diagram for Nd^{+3} ions.

5.1.3 The laser process in Nd^{3+}

The Nd:YAG laser is a four-level process, shown greatly simplified in Fig. 5.2. Level 1, the ground state is the only level thermally occupied. Hence level 2 is not normally occupied, meaning that a population inversion is relatively easy to attain by populating level 3. Of the pump bands from levels 1–4, the strongest are 0.81 μm and 0.75 μm – regions of high output from a krypton lamp. Almost all ions reaching the pump bands are transferred to the upper laser level, and 60% of those radiate at the laser wavelength of 1.0641 μm by falling to level 2.

The laser threshold condition is therefore easy to achieve.

5.1.4 The acousto-optic *Q*-switch

Section 5.1.1 indicated how *Q*-switching can enhance a laser's capabilities. In Fig. 5.1, the *Q*-switch was represented by a 'black box'. What is the *Q*-switch?

A *Q*-switch is a device used to inhibit laser action in a controlled way. It can be a mechanical shutter, a dye capable of being bleached, or a device based on various opto-electronic effects. In most industrial Nd:YAG lasers however, an acousto-optic switch is used. This is a device in which sound waves interact with the laser beam.

Figure 5.3 shows a schematic acousto-optic *Q*-switch. In order to inhibit laser action, sound waves are propagated through the crystal lattice by means of a piezo-electric transducer on one side of the quartz block. An r.f. oscillation (typically 24–27 MHz) generates a series of compressions

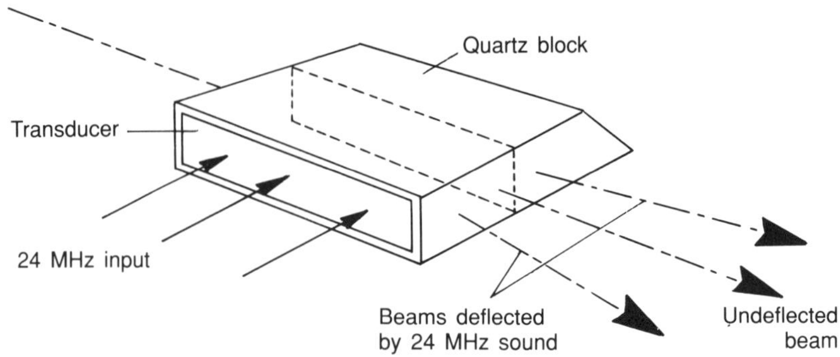

Fig. 5.3 Schematic of an acousto-optic *Q*-switch.

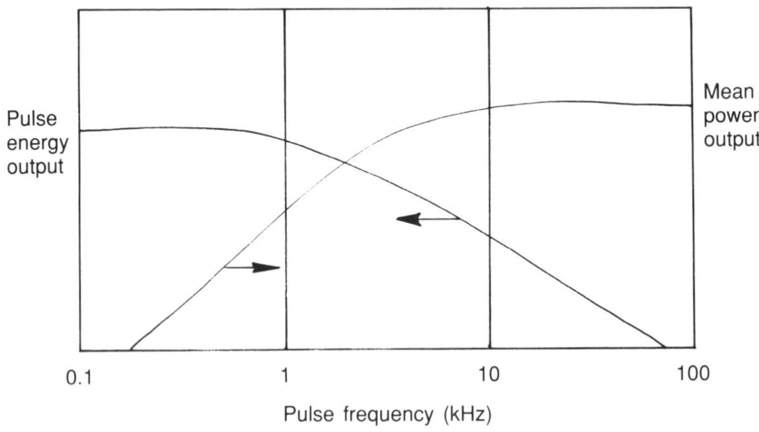

Fig. 5.4 Graph of *Q*-switch frequency versus laser energy and power.

and rarefractions through the material. This periodic adjustment of the refractive index acts as a diffraction grating which, depending on orientation, can create one or more deflected beams.

Provided sufficient r.f. power can be applied, this method can easily deflect sufficient laser power to inhibit the laser action; 60–100 W is typically required to 'hold-off' a 60 W laser beam. These *Q*-switches need to be water cooled. If the r.f. signal to the *Q*-switch is itself modulated (generally at a frequency controllable from 0–50 kHz – the *Q* 'switch frequency'), then the laser will emit a series of energetic pulses at the modulating frequency. Figure 5.4 shows how the energy per pulse and mean power output vary with controlling *Q*-switch frequency. This will be referred to later in the text.

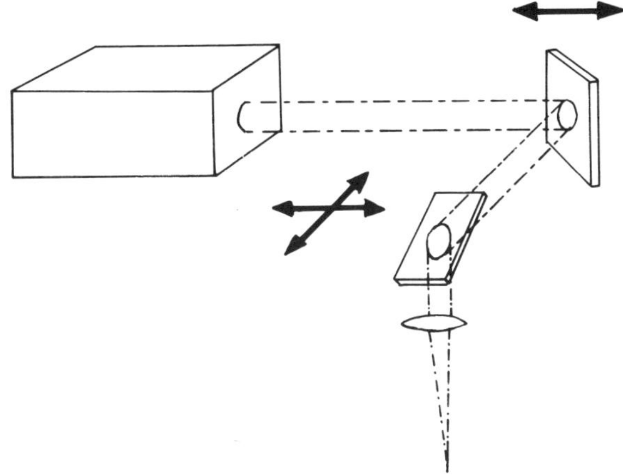

Fig. 5.5 *X-Y* optics.

5.1.5 System formats available

An industrial laser system is likely to comprise: a laser with optics (and *Q*-switch if necessary), parts handling, a control system, and a means of generating movement between the laser beam and the part or worksurface. The last can be generated in a number of ways: move the part, move the laser and optics, or move the optics. A decision on which method to use depends on the sizes of the laser and the parts, the sort of handling required, speed of throughput needed, and cost.

Two popular means of moving the optics only are illustrated in Figs 5.5 and 5.6. The *X-Y* optics in Fig. 5.5 normally need to be mounted on a gantry. Large areas can be accessed and because the mirrors are considerably lighter than the laser, this method is quicker than moving both laser and optics.

Faster still are 'galvanometer-driven optics' shown in Fig. 5.6. Current-controlled galvanometers, controlled by a computer, deflect the laser beam in both *x* and *y* directions. Areas accessible are much less than with *X-Y* optics, but movement is precise (of order 5 μm resolution) and extremely fast (linear movements possible at more than 1000 mms^{-1}), and type-size characters can be described at 60 per second.

The area accessible is limited by depth of field and the size available of precise multi-element optics known as flatfield lenses. A flatfield lens ensures that the beam is focused in a plane perpendicular to the optical axis, regardless of the angle of approach to the lens (i.e. regardless of where in the field is being addressed). Flat marking fields of 160 mm diameter are normal, as much as 200 × 400 mm being available.

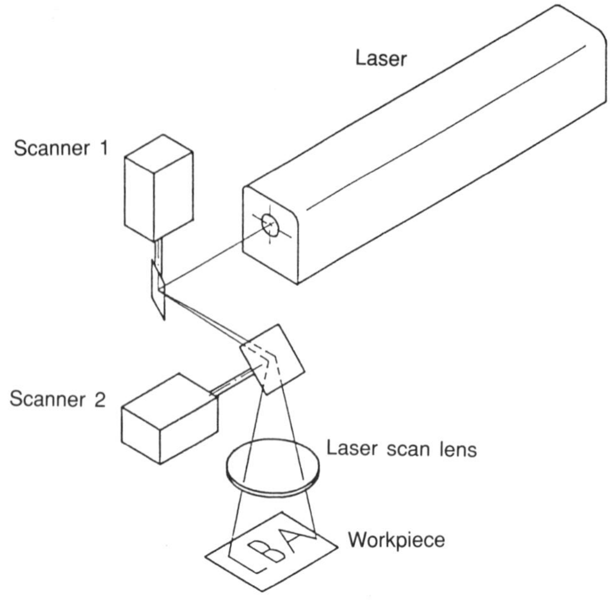

Fig. 5.6 Galvo-optics.

5.2 RESISTOR TRIMMING

5.2.1 Introduction

Many electronic components and circuits are made from thin layers or films of conducting material on an insulating base or substrate. 'Conducting' should be taken in the broadest sense of the word as it can be any material which is not an insulator and in some cases the material can be semi-conducting. Apart from thick and thin film circuits, there are cylindrical resistors coated with carbon film, metal film, metal oxide or thick film, microwave devices made from metal films on ferrites and similar materials, quartz crystal oscillators and filters, surface wave devices and so on. Hardly any of these devices can be made to an exact value and have to be adjusted or 'trimmed' to the necessary accuracy. Laser trimming with a *Q*-switched CW YAG laser trimmer can adjust any of these components and circuits and also can be used for scribing ceramics including alumina and semiconducting materials such as silicon and germanium.

5.2.2 Lasers used in trimming

Laser trimming in general relies on the use of a beam of light to vaporize material very rapidly and at the same time, causing as little damage as possible to the material surrounding this cut. Power densities of the order of 5×10^7 Wcm^{-2} are required for thick film trimming for example, and to achieve this with a continuous output (CW) laser would require output powers of the order of tens of kilowatts. The solution is to use pulsed lasers with high peak power and short pulse length but relatively low mean powers.

The laser pulses are used to drill a succession of holes in the material which overlap to form a continuous cut. By adjusting the peak power per pulse and the degree of overlap, it is possible to remove just the amount of material required, provided, of course, that the available energy per pulse is high enough and that sufficient energy is absorbed to vaporize the material.

Yttrium Aluminium Garnet (YAG) doped with about 0.75% of the rare earth neodymium forms the basis of a reasonably efficient laser system which can be either pulse or continuously pumped to give laser output at 1.06 μm. In pulse operation, the laser is pumped by the light output from a xenon flash lamp. The pulse frequency is limited to a few hundred Hertz and the peak powers achievable are much higher than are required even for thick-film work.

CW operation requires the use of a continuously run source such as a krypton arc discharge lamp. To achieve the pulsed output, a '*Q*-switch' (section 5.1.4) is used which breaks up the output into a series of pulses, and at the same time gives a greatly increased peak power. A brief description of the laser action in YAG is given in sections 5.1.2 and 5.1.3.

Laser trimming consists essentially of vaporizing the material away. It is very important to establish the correct conditions for the best possible trim and it is necessary to optimize the various parameters:

1. Energy or peak power per pulse – controlled by the lamp output and *Q*-switch frequency;
2. Pulse frequency – controlled by the *Q*-switch and its associated pulse generator;
3. Cut width – controlled by the telescope focus;
4. Cutting speed – controlled by the *X-Y* table drive;
5. Multimode or single mode – controlled by an aperture in the laser.

The selection of these various parameters for a particular application depends on a number of factors, but the primary aim is to use the minimum rate of energy that will just complete the cutting action required. If too much power is used, then damage may well occur around the cut which will be determined by the material or device being cut.

5.2.3 Thick-film circuits

Thick-film circuits generally consist of rectangular resistors and conducting pads screen printed and fired on to alumina substrates. The resistor material is a mixture of platinum group metal oxides, a binder and glass powder. The resistors are printed too low in value (usually between –30% and –10%) and increased to value by removing some of the material. The resistance of the films is measured usually as a resistivity in ohms per square – for a given film the resistance across the two opposite sides of a square is always the same no matter what size the squares. The resistivity of the ink is varied by altering the proportion of metal oxide and glass powder. In any one ink system, a range of 1 ohm per square to 1 megohm per square can usually be achieved, although for any given type of ink the range may be smaller than this.

The conductors are made by screen printing metal powders such as silver, gold or platinum and are matched specifically to the thick-film inks to ensure that they can be fired together to give a good bond.

5.2.4 Thin-film circuits

Thin films generally refer to evaporated or sputtered metals or metal alloys on an insulating substrate which is usually glass, but is sometimes alumina. The most-used system is nickel-chromium alloy (nichrome) which has a resistivity of between 100 and 300 ohms per square depending on the film thickness. Evaporated tantalum films are also often used. As the available range of resistivity is much smaller than with thick films, the higher values of resistance have to be obtained by making long narrow resistors which ‘meander’ backwards and forwards across the substrate. These patterns are either photographically masked during evaporation or are cut into sheet material by various trimming methods.

Compared with thick film trimming, much lower peak powers, of the order of a few hundred watts, are needed to evaporate these films. Very fine cuts can be obtained by using the frequency doubled output of the laser. The laser will also cut gold or silver films with a little more power than the minimum needed for nichrome films and can be used to cut the conductors in the circuit. This is particularly useful when parts of the circuit are formed by resistor material bridged with a conductor. Cutting the bridge will then bring the resistor into circuit as required.

5.2.5 Resistor spiralling

Despite the competition from thick-film circuits, there is still an increasing market for discrete resistors, each of which has to be trimmed individually to value. That is except for low-tolerance carbon composition resistors which are being replaced by film resistors of various types.

Film resistors are manufactured from extruded or ground cylindrical alumina blanks which have a film deposited on them. A metal cap is pressed on to each end and wire leads welded on – usually by capacitor discharge.

Sometimes the leads are welded to the caps before capping. The laser spiralling technique involves rotating the resistor at high speed and making a helical cut along the surface by a moving optic system. Because the laser cut can be very narrow, magnification (ratio of final to initial resistance M) can be up to 1000 or more, from the expression:

$$M = \frac{\pi^2 D^2}{P(P - d)}$$

where D = diameter of resistor, P is the pitch of the helix and d is the laser cut width, assuming the whole length is cut.

The actual laser trimming technique is very similar to flat plate trimming for thick or thin film depending on the type of film, except that much higher pulse repetition rates are used.

5.2.6 Scribing

Alumina has a very low absorption at 1.06 μm so that when trimming thick-film circuits, the ink can be removed leaving the alumina substrate more-or-less undamaged. Nevertheless, there is sufficient absorption to drill a hole when a high peak power (10–20 kW) pulse is focused on to the substrate. This can be used to scribe alumina substrates by drilling a line of shallow holes along which it can be broken.

Silicon and germanium can be scribed easily with a YAG laser as they absorb 1.06 μm radiation. Very high speeds up to 200–300 mm per second are theoretically possible with a line width of 12 μm. It is possible to scribe and break all types of devices and laser scribers are used by several manufacturers in USA. The main problem is the redeposition of the evaporated material on to the surrounding devices.

5.2.7 Trimming geometry

Most printed resistors on flat substrates are rectangular with contact pads on two opposite sides or ends. The most basic form of trim is to make a cut from an open edge of the resistor at right angles to the current flow. This is fine provided the change needed is only a few per cent. Calculations will show that the rate of change of resistance with cut distance increases dramatically and as the cut approaches the far side of the resistor the rate will exceed the measuring speed of the controlled bridge. An additional disadvantage is in the narrow current path, leading to overheating and current noise in the final resistor. A technique to overcome this is L-cut, when, after the initial plunge, the cut turns along the current path giving an almost linear rate of change of resistance. Similar results can be achieved by double or multiple cuts from one or both sides and by 'meander cuts' which depend on the geometry (Fig. 5.7).

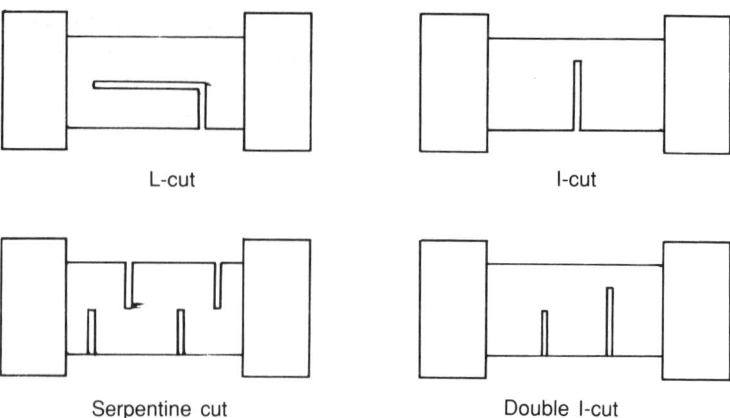

Fig. 5.7 Cut geometries for resistor trimming.

5.2.8 Resistor trimming systems

There are three basic methods of using a laser in resistor trimming:

1. Moving the laser head and optics implies a large mass which is difficult to move quickly between any two points on the substrate. Other problems include mirror vibration and the constant flexing of water and power leads.
2. Moving the optics alone and keeping the laser and substrate fixed is ultimately the best system because the moving mass can be made comparatively small and there are no flexing cables to the laser or resistance measuring probes.
3. Moving the substrate and keeping the laser and optical system fixed means that all the resistance measuring probes and any substrate handling have to move with the substrate. This system does have the important advantage that a simple system with a fixed rugged laser and optical system can be built to be used for development work.

The simplest manual system has the laser head mounted on a pillar which also carries the viewing and cutting optical systems. The pillar is bolted to a base to which any type of *X-Y* table can be bolted under the cutting lens. The whole system is supported at desk-top height for the convenience of the operator. The laser beam is expanded by a Galilean telescope and deflected downwards by a dielectrically coated mirror, before being focused on to the substrate by the cutting lens. The telescope expands the beam to reduce the inherent divergence and hence increase the focal length and working distance of the lens necessary for a specific cut width. A simple resistor trimming system is shown in Fig. 5.8.

The substrate can be viewed through the cutting lens with a television

Fig. 5.8 A simple resistor trimming system.

camera which gives a picture on the TV monitor. This detailed inspection may also be made through a microscope eyepiece, and image inflector, but is possible only with the moving substrate configuration (3 above). This is very valuable in development work and in setting up on production runs.

By adding motorised *X-Y* tables, a resistance measuring bridge and a more complex control box the system can be converted to a semi-automatic trimmer which will automatically trim one resistor on a substrate at a time. The advantages of the semi-automatic system compared with a fast fully-automatic system are the absolute simplicity of programming involving the placing of two probes and setting up a bridge, the absence of expensive probe rings, the higher reliability due to the simplicity of the system and, probably the most important point, the low capital cost so that a number of semi-automatic machines can be used giving the twin advantage of insurance against complete breakdown should a machine fail and the ability to trim different types of substrates simultaneously.

5.2.9 Fully automatic resistor trimmer

The majority of resistor-trimming systems are automatic, using moving optics driven by galvanometer motors, *X-Y* table drives and one or two special systems such as the original ESI moving-coil system. Connections

Fig. 5.9 Trimming station with measuring probes.

to the resistors are by probe rings (Fig. 5.9), with 40 probes or even more connected to a computer programmable bridge by mercury-wetted switches. Solid switches with high enough ON/OFF ratio are now becoming available.

The control computer can be either a mini or micro, with software built in to enable the operator to program quickly all the parameters for trimming relating to the laser, drives, measuring system and resistor geometry. The measuring system can now cope with capacitance, inductance, frequency, and voltage to trim to any parameter and at such a speed that, combined with automatic substrate loading, trim rates of 200,000 per hour and more can be achieved in similar layouts.

The basic trimming methods are very similar to those used in manual systems, but with all the extra equipment to increase speed, improve accuracy and reduce cost.

5.2.10 Conclusion

Resistor trimming was the first major application of lasers in industrial applications and one of the first important uses of lasers overall. Without YAG lasers the electronics industries could well have been still relying on air/abrasive methods of trimming which are slow, dirty, labour intensive and totally unsuited to the latest miniaturized thick- and thin-film circuit hybrids.

5.3 LASER MARKING

Nd:YAG lasers are being used increasingly for marking a variety of materials; for identification, decorative and security purposes. Pure computer control of the nature of the mark means that the process is extremely versatile, clean, fast and cheap to run.

5.3.1 Laser system requirements

Galvo-driven moving optics are generally used (Fig. 5.6). These give speed, precision and an adequate field area for most applications. Typical laser specification is for 60 W c.w. output, Q-switchable up to 30 kHz, and a maximum peak power of 60 kW. Marks with Q-switching have a continuous appearance due to overlapping of spots. Depth of spots is limited by the pulse energy (maximum: 10–20 mJ), but high galvo precision means that marks may be repeated over themselves.

At a working distance of 160 mm, a field diameter of 160 mm is achievable, and writing speeds of 60 type-size characters per second or linear marking of 1000 mms^{-1} is normal. Standard systems offer spot sizes of 25–200 μm by making use of zoom beam expanders, a variety of focal length lenses and apertures to restrict higher-order transverse modes. By frequency-doubling YAG to 532 nm (green) and with special optics, gemstones have been marked with 50 μm high characters, using a spot size of only 8 μm.

5.3.2 Control system requirements

Since different applications will want different styles of marks, the computer should have a library of standard fonts which may be simply called up. Letters, upper and lower case and punctuation/special symbols for a variety of fonts: typeface, italic, script, outline as well as barcodes and machine-readable alphanumerics are available.

Software manipulation allows precise selection of letter size, shape, spacing, linewidth and orientation – for example, writing around arcs or at angles. Graphics permit arrangements of arcs and straight lines into logos, and interfaces to digitizing techniques offer short cuts; in particular CAD links and scanning digitizers. These 'scanners' operate a little like a fax machine; a linear CCD array scans input artwork, diagrams or photographs and generates a raster pattern of binary sequences which is translated into 'laser pulse' or 'no laser pulse' at the workpiece. Digitizing resolutions of around 16 pixels/mm complement laser spot size resolution well. An artificial grey scale can be generated at the cost of somewhat lower resolution. Passport-quality photographs can be generated on appropriate materials.

5.3.3 The effects of laser marking

The Nd:YAG wavelength of 1.06 μm is fairly well absorbed by the whole spectrum of opaque materials. As a general comparison with CO_2 (10.6 μm), the YAG wavelength is rather better absorbed by metals and rather less well by organics. The precise control of average power and peak power offered by a *Q*-switched YAG means that parameters for marking may be tailored to meet the needs of various materials. Different marks may even be produced for a given material; for example, control of peak power enables tool steel to be marked either by engraving or a colour change to black. There follows a discussion of some of the variety of effects of YAG.

Below the melting point, non metals can show a colour change. Metals rarely appear affected, however structural changes are possible. At the melting point many metals show a change in colour and reflectance. Steel, 9 ct gold and brass become dark brown or black. Non-metals often form a groove as material is displaced; changes in surface texture and/or colour change are common.

At vaporization, surface layers can be removed cleanly (for example, paint, ink or anodizing), which can expose the base material giving very high contrast. Most metals and many plastics vaporize giving holes or grooves. Ceramics such as Al_2O_3 (electronics industry) and gemstones also respond.

Chemical transformation of plastics sometimes produces surprising results: dark can change to light or light to dark. Careful control of filler materials used (e.g. titanium dioxide, mica) can allow useful manipulation.

Transparent materials do not generally mark. This can be an advantage as marking can be performed through glass covers (for example on the face of a sealed electricity meter). Translucent materials usually present problems as absorption is not controlled on the surface, but throughout the body of the material. (Example: silicon wafers in section 5.3.5.)

The general criteria for selection of the correct *Q*-switched frequency and power setting to produce the type of mark required, are that a low *Q*-switch frequency produces a low average power and high peak power – good for vaporization and engraving. A high *Q*-switch frequency gives a high average power and low peak power (quasi cw in the extreme), which is good for thermal effects: oxidation, annealing, colour change, melting (Fig. 5.4).

5.3.4 Advantages

The non-contact nature of laser marking means that materials experience minimal deformation or microcracking, tool wear is zero, and consumables are therefore low. Unlike stamping, whether by ink or indentation, fixing of components with clamps or jig is largely unnecessary. Inaccessible, curved or fragile surfaces may be marked. Speed and flexibility mean that on-line marking is possible. Computer generation of marks makes integration with other devices such as handlers or barcode readers easy. Uniform characters and fine quality logos generally mean that finished products may be marked.

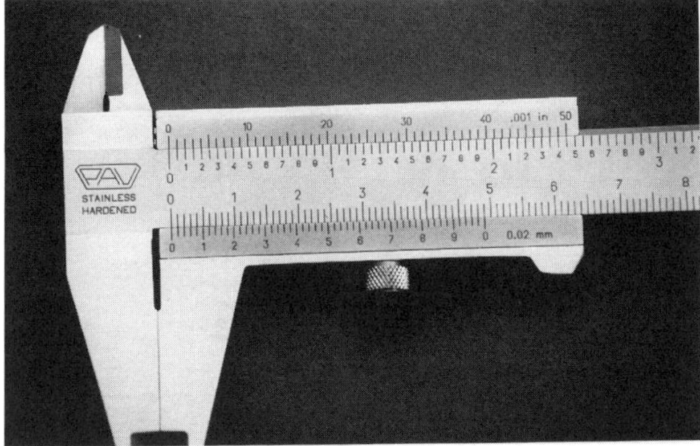

Fig. 5.10 Laser-marked vernier calipers.

Fig. 5.11 Label material specially developed for marking by laser.

5.3.5 APPLICATIONS

The variety of applications of these processes is enormous. Three particular ones will be highlighted in this section; the many others include: vernier calipers (Fig. 5.10), gauges and thimbles (scales marked up from blanks); labels made of plastic laminates (Fig. 5.11) or anodized aluminium

Fig. 5.12 Some samples of barcodes marked by laser.

Fig. 5.13 Decorative marking on a sword blade.

(Fig. 5.12); reactor and aircraft components requiring permanent traceability; decorative and personalized marks on pens, or for example a sword blade (Fig. 5.13); electronic components; radio and car dash logos and scales on buttons and knobs; typewriter keyboards; ball-bearing races (Fig. 5.14); turbine blades.

Hard metal tools

Carbide tip inserts, cutting tools, drill bits, taps and dies and cutlery all come under this category. Conventional methods tend to be either

Fig. 5.14 A laser-marked bearing race.

electrochemical etch (which is labour intensive and generates unreliable, mediocre quality marks), ink screen print (wears off quickly) or stamping (inflexible and in the case of hardened tools, expensive and damaging).

Tools can be bright, in which case a black, annealed colour-change mark is preferred, or chemically-blackened when a bright mark is extremely visible and quick.

High production numbers mean that the laser capability of a serialized mark on each part in a time of the order of one tenth of a second easily justifies the capital outlay.

Surgical implants

Special hard alloys tend to be used, for example, for hip joints. Their irregular shape, restricted surface area designated for identification and extreme requirement for traceability in the event of failure, make them an obvious choice for laser marking. In addition, any mark on units to be implanted must be totally non-toxic.

Silicon wafers

When raw silicon wafers pass through the many stages of fabrication to become microchips, they need tracing. Improvement of yield is critically dependant upon knowing which batches or production procedures generated faulty units. Devices with tracks and insulation of micron dimensions need close dust-particle control; a marking process must be clean – often to Class 10 requirements. A mark should therefore not be ablative, but a shallow melt mark.

Silicon wafers are single crystals; this means that damage on an atomic scale can, in principle, pass through the entire crystal lattice by means of dislocation. Every effort needs to be made that marks are extremely gentle and restricted to the surface. To compound the problems, marks should take up the minimum of useful space possible and need to be machine readable. YAG lasers are performing this task at nearly every semiconductor factory in the world. Extremely careful control of laser power is exercised, so that the minimum energy required is used. This means that despite the partial transparency of silicon to infra-red wavelengths, largely surface marks are achieved.

A recent development to enable a better safety margin for the process has been to use frequency-doubled YAG radiation. Silicon absorbs green light rather better, and a less damaging, more superficial mark is (relatively) easy to achieve. Being of a shorter wavelength, 532 nm, green light is focusable to a smaller spot. It is this which has made possible the traceability of individual chips by marking each one using such a laser system.

5.4 LASER SOLDERING

Laser soldering is not predicted to replace conventional methods, namely wave, infra-red or vapour-phase soldering. Those are parallel methods which are therefore quick, and they represent acceptable investment levels for production of routine boards. There are, however, certain aspects of laser soldering which mean that 'serial' soldering (or at best two at a time) methods are still attractive. These are discussed in sections 5.4.1–5.4.4.

5.4.1 General requirements of laser soldering

During the early stages of the development of a laser soldering system, capable of being incorporated within an SMD (surface mount device)-based circuit production line (Fig. 5.15), a number of sensible broad requirements are laid down. It should be able to solder existing surface mounted components, (e.g. Fig. 5.16) and those predicted for the future. A maximum size of 40 × 40 mm was conceived. The board is expected to be stationary while a given component is being soldered, but larger boards would certainly require 'stepping' from one area or component to another with an *X-Y* table.

Soldering should be performed using two laser beams operating simultaneously. This avoids an effect known as the 'tombstone' or 'Manhattan' effect – the springing-up of lighter components on end when a single point is soldered. Both laser beams should be independently controllable in terms of both position and energy applied per pad. Figure 5.17

Fig. 5.15 A production system designed for in-line SMD soldering.

Fig. 5.16 A laser-soldered component.

Fig. 5.17 Soldering area of a production system.

Table 5.1 Energy absorbtion percentages for CO_2 and Nd:YAG laser

Laser	Material	
	Tin/lead solder (%)	Composite P.C. Board (%)
CO_2 (10.6 μm)	26	100
Nd:YAG (1.06 μm)	79	73

shows the soldering area of such a system with independent galvo control on both beams. Finally, the installation should be completely CAD/CAM compatible to permit a viable link on a production line.

5.4.2 Choice of laser

Two well-proven industrial lasers are worth consideration for the job of automatic laser soldering: CO_2, operating at 10.6 μm and Nd:YAG, at 1.06 μm. Table 5.1 shows the energy absorption percentages for each of these wavelengths on solder and pc board composite. With a target in mind of

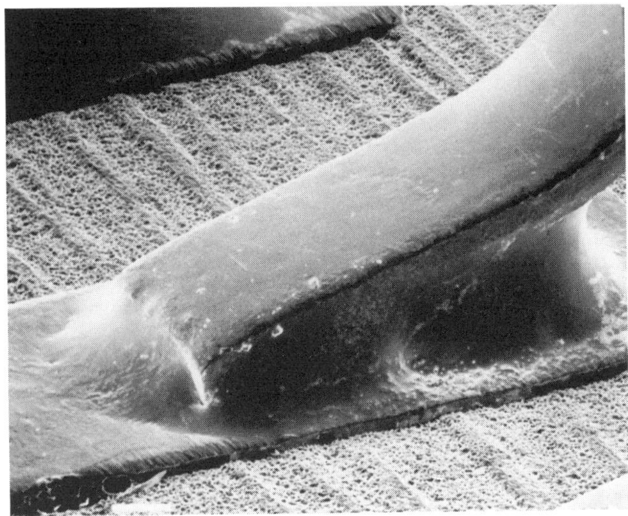

Fig. 5.18 Electron microscope picture of a 'wetted' joint.

coupling energy efficiently into solder material with the least damage to the board itself, clearly Nd:YAG is the better option.

Two broad catagories of YAG are available: pulsed YAG, where the pumping flashlamp is pulsed, and cw (continuous wave) pumped YAG (which is commonly *Q*-switched). Pulse durations from pulsed YAG lasers are typically in the range of a few milliseconds down to tenths of milliseconds. On this sort of timescale, the power applied is too high; solder tends to vaporize and throw up small quantities of solder – 'solderballing'. Clearly this is not an attractive property, and the steady melting supplied by a cw YAG is much preferred.

For small joints a cw YAG laser beam of 20 W will reflow solder paste in 25–40 ms. Greater than 20 W input power can generate solderballing and should be avoided. Figure 5.18 is a scanning electron micrograph of a successful soldered joint.

5.4.3 System configuration for fully automatic soldering

The main components should be integrated into a laser safety Class 1 housing. These comprise:

Laser, optics and beam handling

20 W per beam may be extracted from a cw Nd:YAG laser nominally rated at 60 W. Beam splitters may be used or, more elegantly, output can

be taken from *both* mirrors in the laser, i.e. both are made partially transmitting. Each of these beams should be independently controllable using a 'galvo head' with a pair of scanning mirrors (section 5.1.5 and Fig. 5.6). The galvo heads should be mounted at an angle to the vertical (commonly 30°) to ensure clear access to solder points on, for example, leadless chip carriers. A low-power red helium neon laser beam is often mirror injected so that it follows the same optical path as the YAG beam. This serves as a visible pointer and aids set-up.

Control and system management

A powerful computer controls the laser and galvo optics; a hard-disk memory is required to satisfy the large 'library' of instructions necessary. The software 'library' can reference any component likely to be used and supply, for each, the pad co-ordinates, the required beam geometry of movement for each, and the laser energy per pad. The computer must also interface to the CAD instructions on the preceding pick-and-place machine, receiving from it a full description of orientation and position for each component.

Automatic transfer systems and implementation into production

The complete production line begins with blank boards which are screen printed with solder paste and ink as necessary. Each board is conveyed automatically to the CAD-driven pick-and-place unit which applies the surface-mounted (i.e. no 'through' wires) components to the solder paste pads. It also tells the laser soldering system the position of each component. The conveyor to the laser soldering system should have an input buffer of boards to be soldered, and a carrier to take each board into the soldering position. Here, an *X-Y* table steps the board so that the laser beams can access any component. An output buffer then feeds completed boards into clean test areas.

As an example of the operator's viewpoint, Fig. 5.19 shows a soldering 'event' as seen on the built-in CCTV monitor.

5.4.4 Benefits of laser soldering

Using laser soldering, heat can be applied precisely to the pad. This means that both the board and components remain cool, and the process can be used for thermally sensitive components. Additionally, thermal stressing of the board is avoided. The joints produced by laser techniques are of high and reliable quality. They are repeatable because of the controlled, programmed nature of the process and the way it applies individually tailored parameters to each joint. The joint quality is good because rapid

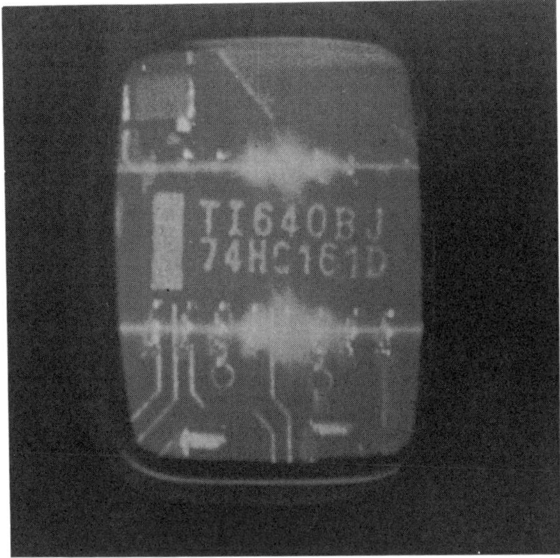

Fig. 5.19 CCTV picture of soldering event.

heating, then solder flow followed by rapid cooling (because so little energy has been used), means that the intermetallic zone is small and the metal grain structure is fine. These are characteristics of joints with good conductivity and high mechanical strength.

The process lends itself well to prototyping. Once the pick-and-place machine is programmed as to what components to put where, all that is required is the solder system library and a CAD download. Empirical cycling of temperature and time periods is not required. The time to produce a single unique board need be no longer than full-scale batch production of identical units. The process is the fastest serial soldering method available, being capable of up to 30 joints per second. It also generates less toxic material and waste than most competitive techniques.

Figures 5.16–5.19 illustrate some of the above points.

5.5 A PERSPECTIVE

This chapter has taken a practical viewpoint of just three broad applications of one type of laser – the cw pumped neodymium-doped YAG laser. There are many other applications for this laser such as drilling, cutting and scribing of metal foils, diamonds, alumina and some plastics. Diamonds can have impurities ablated, integrated circuits may have 'shorts' removed, gyros can be balanced by mass removal and quartz filters can be adjusted.

Neodymium doped YAG can also be pulse pumped with a flash lamp to give a much greater energy for larger scale metal removal for cutting and drilling, and the longer pulses allow metal welding. New high power 1 kW continuous YAG lasers have considerable advantages over similar power CO_2 lasers due to the superior absorption of metals at 1.06 μm.

For the future, compact, lightweight, ultra-efficient diode laser pumped YAG lasers will become the norm, utilizing not kilowatts of electricity and many gallons of water for cooling, but hundreds of Watts and air cooling. Nd:YAG is just one lasing medium amongst many including CO_2 and CO. Ruby, Argon Ion, Helium-Neon and Helium-Cadmium lasers each have their own application areas. New advances are made every year in the field of laser diodes which have exceeded 1 W output and are moving to the visible, and now the excimer laser, with its UV output, makes micromachining of materials possible at dimensions hitherto considered impossible.

6

CO_2 gas lasers: engineering and operation

M.J. Adams

6.1 INTRODUCTION

Previous chapters have already introduced the physics of lasers, the principles underlying their design and some of their many applications. This chapter deals with the design and development process which turns those design principles into a laser working in production. This is a fascinating and very often frustrating process which has occupied the author for many years.

Lasers are fascinating because they are very subtle machines and therefore present a considerable challenge to a designer. A laser, to look at, is a deceptively simple device. A gas laser, for example, consists simply of a gas-filled tube with mirrors at the end, and of course the odd vacuum pump and cooling fan. To make a gas laser 'lase' is also simple, but to make it produce a stable, usable mode for hour-after-hour and week-after-week in many different industrial environments, is far from simple. Lasers are frustrating because, being subtle, they are very difficult to persuade to operate as the designer wishes.

Consider the kind of engineer best qualified to design and develop lasers. He or she is probably not a physicist, although physicists are often chosen to do the job. In fact a laser designer is probably best described as a Jack-of-all-trades, as he must have some knowledge of physics but also know something of high-voltage electricity, mains electricity, electronics, vacuum techniques, optics, control systems, pneumatics, and most important of all, materials science. In the past, designing lasers was often the work of one person but as the industry matures and competition strengthens it is increasingly likely to be carried out by a team, in which each member specializes in a certain area.

It is important that the leader of the team is able to look at the task

Laser Processing in Manufacturing. Edited by R.C. Crafer and P.J. Oakley.
Published in 1993 by Chapman & Hall, London. ISBN 0 412 41520 8

through the eyes of many people. Firstly, he or she must be enough of a physicist to be able to ensure that the basic design principles are sound. Certainly the performance of the laser must be good, but that is only the beginning. It must also be very reliable, cost effective, easy to install, operate, service, maintain and repair, and also similar or lower in running costs to any competitor's equipment. It must also be safe.

Thus the leader of a laser design team must also be able to look at the design through the eyes of a service engineer, a safety officer and very importantly, through the eyes of the customer.

It is not sufficient for such a person to be a good administrator because he or she must also have sufficient technical knowledge of a wide range of subjects to be able to guide the other team members in their specialist areas. This can be said of any team leader but in the case of lasers the range of knowledge required is perhaps wider than usual.

6.2 IMPORTANT DESIGN PARAMETERS AND CONSIDERATIONS

6.2.1 The required power output

This is usually the most important single factor influencing the design of a gas laser. The CO_2 laser is one of the most efficient types of laser but its overall or wallplug efficiency is still only about 10%. (It is interesting to note that this inefficiency is compensated for by the laser's precision. A plasma arc or a mechanical saw will make much wider cuts than a laser and so must remove much more material. Thus although these processes are much more efficient than a laser at converting electricity, the power required for them to do the cutting is quite as much as that needed by a laser.) All lasers, therefore, produce unwanted heat, and as the power level of the laser increases, the problem of getting rid of this heat increases also. Since the unwanted heat is generated in the laser gas, the problem is one of cooling this gas, and as the required laser power increases, more effective means of gas cooling must be introduced by the laser designer. The chosen method of cooling the laser gas may itself influence the electrical excitation method, and therefore the type of power supply employed, since the two processes are not independent.

Low-power, sealed or no-flow lasers

Low-power lasers, that is lasers with cw power outputs up to say 60 W, are used for medical applications such as surgery and for some industrial applications which involve cutting and welding thin, non-metallic materials.

Examples of these applications are ceramic substrate scribing, perforating thin polythene used to make air beds for burns patients in hospital, simultaneously cutting and welding polymer sheet to form special purpose packaging, cutting fabric to make boat sails and drilling holes in babies' bottle teats and aerosol nozzles. Work-handling systems for many of these applications are simplified if the laser head has few service connections to its power supply, is small, and is light in weight.

When the power required from a CO_2 laser is around 50 W or less, it is relatively easy to remove waste heat from the lasing gas. In addition, contaminants are formed in the gas discharge at a relatively slow rate when the laser is operating. By adding extra ingredients to the normal lasing gas mixture of carbon dioxide, nitrogen, and helium and paying great attention to the cleanliness of the system, the designer is able produce a completely sealed laser which will run for long periods, in some cases thousands of hours, before the plasma tube with its charge of gas must be replaced.

When electrical discharges are operated with electrodes in contact with the gas, electrons and ions bombard the electrodes and remove small amounts of material by a process known as sputtering. In sealed lasers the sputtered material contaminates the laser, reducing it's power. For this reason sealed lasers use radio frequency (RF) excitation, which may be operated with electrodes placed outside quartz or ceramic plasma tubes. Electrical power is coupled capacitatively into the discharge through the tube walls. Since the electrodes are transverse to the plasma tube axis, and typically only millimetres apart, the exciting voltage is relatively low (a few kilovolts). Spacings between components need not be large to allow for electrical flashover and tracking, and the laser head can be smaller than if the excitation were DC or high frequency (HF).

Thus RF excitation has two advantages when applied to low-power CO_2 lasers: it allows them to use a sealed gas system and it enables the size (and therefore the weight) to be reduced. In some sealed lasers the size may be reduced even further by using very small diameter plasma tubes which are reflective on the internal surfaces and so form an active part of the resonator. Such devices are known as waveguide lasers.

Slow-flow lasers

As power levels increase to 100 W cw power output and beyond, it becomes more and more difficult to remove excess heat from the laser head and to maintain stable beam conditions. This is especially difficult if the laser is small in size. In order to cool and stabilize the lasers effectively they must be made somewhat larger. If the lasers are larger, there is less reason to use RF excitation which, as stated above, is in any case appreciably more costly and less efficient than DC. Thus most CO_2 lasers of 100 W power output and more are DC excited.

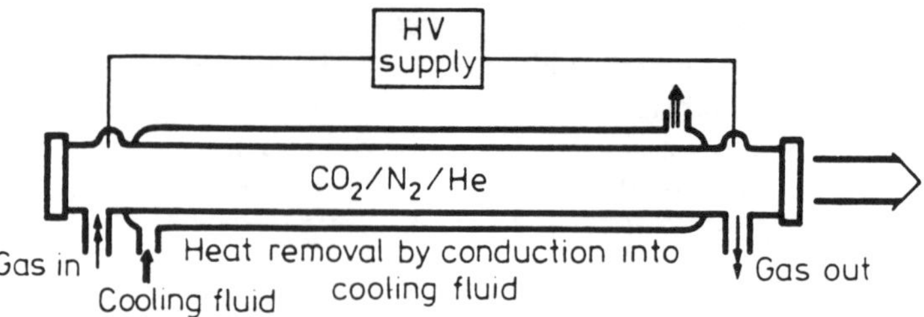

Fig. 6.1 Slow-flow CO_2 plasma tube showing gas connections and water jacket.

Furthermore, as power increases, impurities are produced at higher rates in the discharges and it becomes much more difficult to design a device which is completely sealed off. For this reason, most lasers of 100 W and more are of the slow axial flow type where the lasing gas is fed from high-pressure cylinders into the plasma tubes and flows slowly through them at a few metres per second (Fig. 6.1). Slow axial flow CO_2 lasers are relatively simple to design, build and maintain, and predominate in the power range 100–400 W cw power output.

However, as always in laser design, a problem is to remove waste heat from the discharges. It may be shown that the temperature of, and thus the maximum power obtainable from the tube is dependent only on the length of the excited region of the plasma tube (Chapter 1).

This also applies in the case of sealed lasers, which share the same cooling mechanisms. Thus the output power of slow-flow and sealed lasers depends only on the total length of the discharges and cannot be increased by increasing the tube diameter. In practice it is possible to obtain about 60 W per metre of discharge length so that a 300 W laser would need about 5 m of discharge tube. Since in practice industrial customers appear to be willing to accept a maximum length of about 3 m for a CO_2 laser, it is necessary to reduce the overall length of any industrial CO_2 laser of more than about 150 W output by folding it optically with mirrors.

Optical folding greatly increases the mechanical complexity and the difficulty in obtaining good beam quality, especially when more than one fold is used (Fig. 6.2). It is generally preferable to limit the number of folding mirrors to a maximum of two to give an acceptable beam quality, although it must be said that good quality beams have been obtained from some multi-mirror designs. However, lasers using many mirrors are necessarily more difficult to maintain and the time taken for realignment after cleaning optics can be considerable.

The fast axial flow CO_2 laser

In the fast axial flow (FAF) CO_2 laser the gas mixture is forced through the plasma tubes at velocities up to 400 m per second. It is recirculated

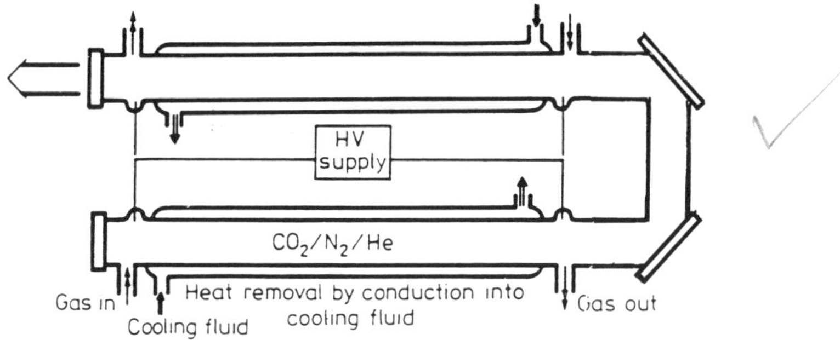

Fig. 6.2 Slow-flow CO_2 laser with one optical fold to reduce overall length.

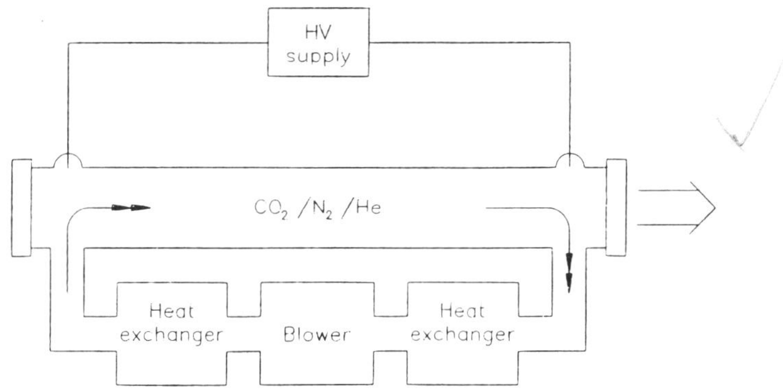

Fig. 6.3 Schematic of fast axial flow CO_2 laser.

through the tubes many times and after each passage through the discharge is cooled directly in a gas-to-liquid heat exchanger (Fig. 6.3). The laser's power output is no longer dependent on the length of the discharge tubes but largely on the rate of mass flow of gas through them. Laser power can therefore be increased either by increasing the volume flowrate or the pressure of the recirculating gas, provided that there are no other mechanical or electrical limitations. In practice the power output of a FAF CO_2 laser may be limited by one or more of the following factors:

1. Volume flowrate of the gas recirculating pump;
2. pressure drop across the gas recirculating pump;
3. operating pressure of the laser;
4. voltage available from the power supply;
5. current available from the power supply.

As a simple rule of thumb the power output of a FAF CO_2 laser should be around 1.5 W for each cubic metre per hour of pumping capacity. Thus, if the power supply is sufficiently large a laser fitted with a 1000 m^3/h recirculating pump should produce approximately 1500 W.

Although it may appear that power could be increased indefinitely simply by increasing the operating pressure of the laser, this is not the case. Unfortunately, as running pressure rises the tendency of the discharge to form an arc rather than a homogeneous glow also increases and in practice few FAF CO_2 lasers operate above 150 mbar pressure.

Since the pressure drop around the gas recirculating system is proportional to the pressure at the input to the recirculating pump, any rise in the system operating pressure will cause the pump to work harder and raise it's operating temperature. This will generally reduce the lifetime of the pump and if pressure is raised too far, may overheat it and cause the rotors to contact the casing, so that the pump rapidly seizes up.

It may also appear to be a simple matter to increase power by using a larger capacity recirculating pump, but again a practical limit is reached when the plasma tube 'chokes'. This occurs when the gas velocity in the tube reaches the local speed of sound in the lasing gas mixture. As volume flow is further increased the gas velocity remains constant and the pressure in the tube begins to increase so that the power output is again limited by the tendency for the discharge to form an arc.

At this point the only way of obtaining more power is to increase either the plasma tube diameter, or the number of plasma tubes used in the laser, so that a larger volume pump may be used without increasing the pressure or pressure drop in the system. Each approach has its problems.

If the plasma tube diameter is increased too much for a given resonator length, the laser may operate in a higher-order mode or with a higher beam divergence, which may reduce the focusing properties and thus the material processing performance of the laser even though the power has been raised. If, on the other hand, the number of plasma tubes is increased, the complexity and therefore the building cost of the laser will also increase. This may also give rise to problems of physical size and also of reliability, since more components usually means poorer reliability.

For reasons connected with the symmetry of the gas flow system and the simplicity of the electrical system, it is not usually desirable to add just one extra plasma tube in order to increase the power of a laser – lasers are rarely seen which make use of an odd number of tubes. In addition, designers usually prefer to develop a range of lasers based on a modular system to limit the total number of components employed. Such a system usually involves a basic module of two or even four plasma tubes; for example, the Electrox fast-flow laser range at present contains two-, four- and eight-tube models. Much time has been spent in minimizing the total number of components used and even the lasers at the extreme ends of the this range, a 300 W and a 2000 W, still contain a very large percentage of common components.

Thus, to design a fast-flow laser of a given power output it is necessary to choose carefully the size and type of power supply and recirculating

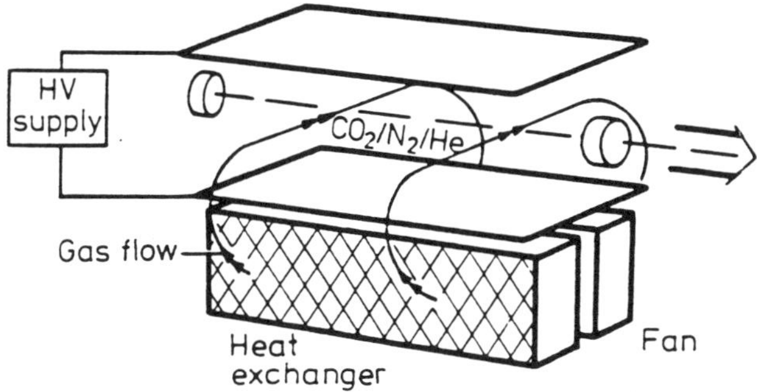

Fig. 6.4 Schematic of transverse-flow CO_2 laser.

pump required. These must also be appropriate for the design of resonator which, for reasons of economy, should use the minimum number of plasma tubes needed to give the required power in a high-quality mode to ensure good process performance. The resonator design is discussed in some detail below.

The fast transverse flow CO_2 laser

As required power output levels rise above around 5 kW, the design of fast axial flow CO_2 lasers becomes increasingly difficult. The power required to recirculate the lasing gas is considerable. For example, the 1000 m^3/h pump used in a 1500 W laser needs a 4 kW motor to drive it; for a 6 kW laser the figure will be at least 15 kW. In order to obtain sufficient mass flow of gas through a 6 kW FAF laser it will be necessary either to use a large number of plasma tubes, perhaps 12 or 16, or to use large diameter tubes, when it may be difficult to control the output mode.

If small tubes are used in order to obtain a high-quality mode, an additional and very serious problem arises – thermal lensing of the output window will occur. This causes distortion of the beam and an effective increase in the divergence, which adversely affects the focusing properties of the beam. This in turn will reduce the effectiveness of the laser in many applications.

For these reasons most lasers above 5 kW power output are of the transverse gas flow type in which the lasing gas flows in a channel at right angles to the resonator axis. In most cases the discharge is between electrodes which are also at right angles, both to the resonator axis and to the direction of gas flow through the discharge region (Fig. 6.4). In such lasers the area of the gas flow path can be made much larger than that in axial-flow lasers. Consequently the flow resistance and thus the power required to drive the gas recirculating pump is much reduced. Despite their advantage of reduced pumping power, transverse-flow CO_2 lasers do suffer significant disadvantages compared with axial-flow types, which is why the latter are more generally used at lower power levels. In an axial-flow laser the mode is to a large extent controlled by the plasma tube diameter, which may be designed to act as an effective optical aperture. No similar control is available in a transverse-flow laser so that it is more difficult to produce a mode of comparable quality. Mode control is made even more difficult by two further factors.

Firstly, there is a tendency for the discharge to be displaced downstream by the gas velocity so that the excited gas or plasma does not coincide exactly with the mode volume. Secondly, transverse-flow lasers usually contain several folding mirrors to reduce the physical length while allowing the resonator to be long enough to produce a reasonably low order mode. The discharge displacement tends to reduce the electrical efficiency of the transverse flow laser, and the number of optics tends to introduce distortions into the beam. However, in spite of this, transverse flow is usually the preferred solution for the design of CO_2 lasers with power outputs above 6 or 7 kW.

When a transverse excitation design is employed, the excited volume must be persuaded to extend evenly across the full width of the gas flow channel. This can be done in two ways. If the laser uses DC excitation the cathode must be divided into a large number of separately ballasted segments, so that the current will divide fairly evenly between them. Alternatively the discharge may be RF excited. In this case only two electrodes are necessary and the electrical supply is greatly simplified at the laser head, but at the expense of using a larger and more complex power supply which is also relatively inefficient. In practice, both DC and RF excitation is used for lasers at power levels of 5 kW and upwards and both methods have their adherents.

6.2.2 Types of power supply

At present, all industrial CO_2 lasers are electrically excited. Electrical power is coupled into the lasing gas, a mixture of CO_2, N_2 and He, to form, ideally, a homogeneous glow discharge completely filling the mode volume of the resonator. The discharge may be excited by a power supply which supplies either DC current or AC current which alternates at frequencies between that of the mains and perhaps as much as 100 kHz.

DC power supplies

The DC power supply is very simple and consists essentially of a high-voltage transformer and rectifier with a large smoothing capacitor. The power supply must also incorporate some means of stabilizing the dis-

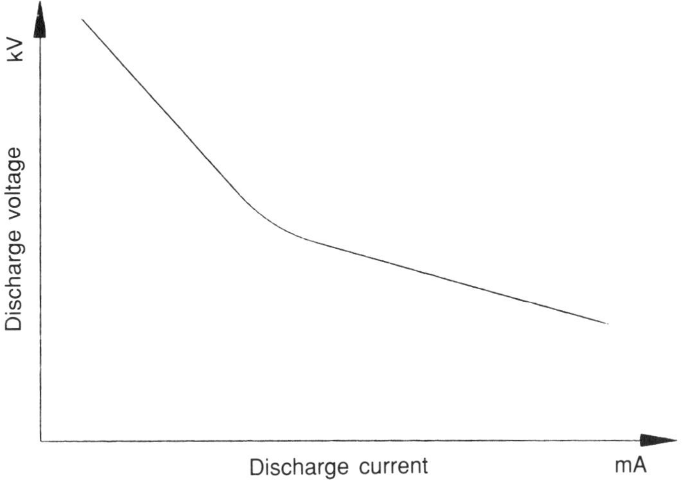

Fig. 6.5 Voltage–current characteristic of slow-flow discharge.

charges since a gas discharge of this type has a negative voltage/current characteristic. This means that, unlike a resistor, for example, the voltage across a CO_2 laser discharge falls as the current through it rises (Fig. 6.5). Thus, without some additional means of stabilization, the current is very difficult to control and if, for example, the voltage is reduced in an effort to reduce current, very little will happen until the voltage is too low to sustain a discharge at all, when it will extinguish. This effect is counteracted by placing a device with a more strongly positive discharge characteristic in series with each plasma tube. The device may be a resistance, an inductance or an active element such as an electron tube which, as well as stabilizing the discharge, may also be used to control the current level and to switch the discharge on and off very quickly in order to pulse the laser. A typical DC-powered CO_2 laser is shown in Fig. 6.6.

High-frequency (HF) power supplies

In an HF or high-frequency power supply the current alternates at a frequency of between about 20 and 50 kHz. Such a supply has two advantages over conventional DC supplies. In an HF supply the AC power is generated at relatively low voltage and then transformed up to the laser operating voltage of between 10 and 30 kV. The transformers are considerably smaller and lighter in weight than the mains-frequency transformers used in conventional DC lasers because the amount of copper in the windings is reduced as the operating frequency is increased. HF power supplies also use solid-state control systems. This obviates the need for the

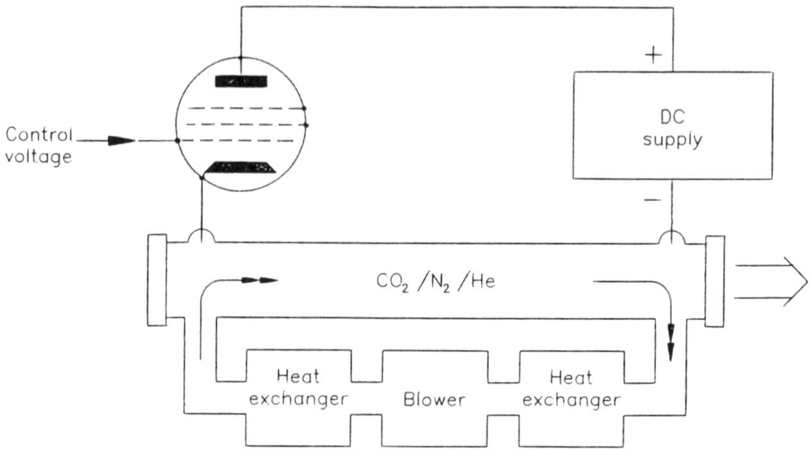

Fig. 6.6 Fast flow laser with DC power supply using electron tubes to control discharge current.

electron tubes used for current control in DC supplies and gives the HF power supply a slight advantage in overall efficiency.

Before being applied to the plasma tubes the HF voltage is normally rectified and smoothed, so that as far as the operation of the discharge tubes is concerned, the HF power supply is identical to the DC supply and in this respect the two can be considered equivalent.

Radio-frequency (RF) power supplies

An RF electrical supply for a CO_2 laser is one in which the current alternates at a frequency of between 2 and 100 MHz. Firstly, the RF signal is generated at the appropriate frequency by a quartz crystal oscillator circuit and then amplified to the required power levels needed to excite a CO_2 laser. RF power supplies were first used in CO_2 lasers in an attempt to overcome the problems of arcing and electrode wear in DC-excited lasers, and in this respect they have been very successful.

The arcing problem does not exist with RF excitation because the current flow period is much less than the millisecond required for an arc to develop. Also, there is very little problem with electrode wear, since, at RF frequencies, the power may be coupled into the discharge capacitatively through electrodes placed outside the plasma tubes. Thus even if some electrode wear takes place, it will not produce dust inside the laser. RF excitation also uses lower voltages than DC because the electrodes are placed at each side of the discharge tube (Fig. 6.7). A typical exciting voltage is 4 kV.

However, the actual generation of RF power is a relatively inefficient

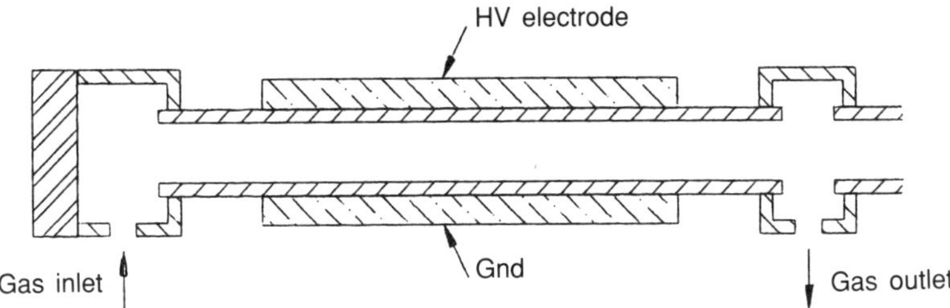

Fig. 6.7 Schematic of RF-excited plasma tube with external electrode.

process. The RF signal is usually amplified in several stages; at the high power levels needed to excite a CO_2 laser, the amplification is carried out using electron tubes. The efficiency of the amplification process seldom exceeds 50%, so that an RF-powered CO_2 laser has an overall efficiency much lower than that of a DC laser. In addition, the RF power supply is much more complex and costly than the DC equivalent. In addition, it is difficult to make the discharge completely fill the plasma tube volume, which results in a further reduction in efficiency. However, this is offset to some extent by the fact that the RF discharge has no cathode fall region, and also uses less power to recirculate the laser gas.

6.2.3 Pulsing capability

CO_2 lasers may be operated to give a true continuous DC(cw) power output if they are DC excited. They may be pulsed simply by switching off the plasma tube current at the required pulse frequency. HF- or RF-excited lasers also give a cw output although the current input to the plasma tubes is pulsed, at a few tens of kilocycles in the HF case and a few megacycles with RF. Such lasers however give essentially a DC output because the lifetimes of excited molecules in the plasma are long compared with the HF or RF frequency, so that the plasma does not extinguish between pulses.

Pulsing DC CO_2 lasers

The plasma tube current can be pulsed using thyratrons to switch the primary current of the high-voltage transformer which supplies current to the plasma tubes. This method is limited to rather low frequencies, usually well below 1 kHz, because of the inductance of the transformer windings. It does have the advantage, however, of switching current at mains voltage rather than at the relatively high voltage needed to operate the plasma tubes.

A much faster and more flexible way of pulsing DC lasers is to use an

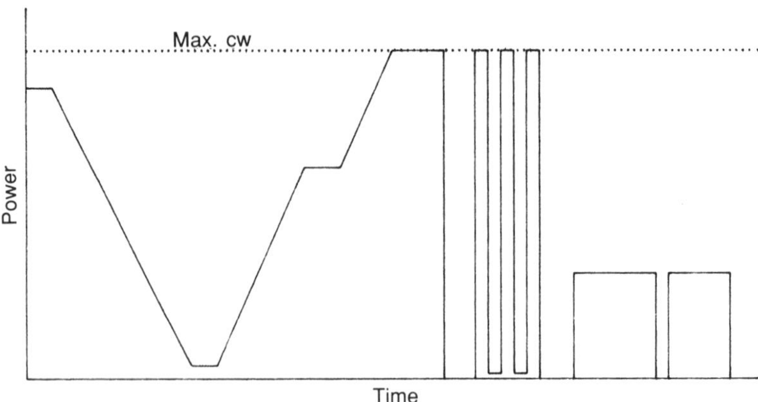

Fig. 6.8 Pulsing of DC-excited CO_2 laser.

electron tube, usually a tetrode or triode, which is placed in series with the cathode of each plasma tube. In operation, a signal is sent to the grid of each electron tube and is used to switch the plasma tube current, and thus the laser power, on and off at any desired frequency. Maximum pulse frequency is limited only by the speed of response of the discharges. At frequencies up to about 4 or 5 kHz the normal cw gas mixture may be used and the peak power of the pulses is equal to the cw level.

To pulse at higher frequencies the amount of CO_2 and especially nitrogen must be reduced in order to increase the rate at which the laser output responds when the tube current is switched off. As the gas mixture is changed to increase the effective pulse frequency the cw power level is progressively reduced. With the gas mixture required to pulse at frequencies of 10 or 12 kHz the cw power may be reduced to 10% of its normal value. These limitations are imposed by the physics of the gas discharge and will be similar whether the laser is DC, HF or RF excited. DC lasers controlled by vacuum tubes are particularly flexible as regards pulsing since they may be operated at any combination of cw power level, mark-to-space ratio and frequency in order to obtain the best output for a particular application (Fig. 6.8).

Pulsing HF CO_2 lasers

High frequency (HF), or switch-mode power supplies for CO_2 lasers operate at an AC frequency of a few tens of kilocycles. They are usually pulsed by switching off the AC current for an integral number of cycles and then switching it on again. In practice this means they can be pulsed at frequencies up to 1 kHz or perhaps as high as 2 kHz. This means that they are suitable for applications such as metal cutting and hole drilling

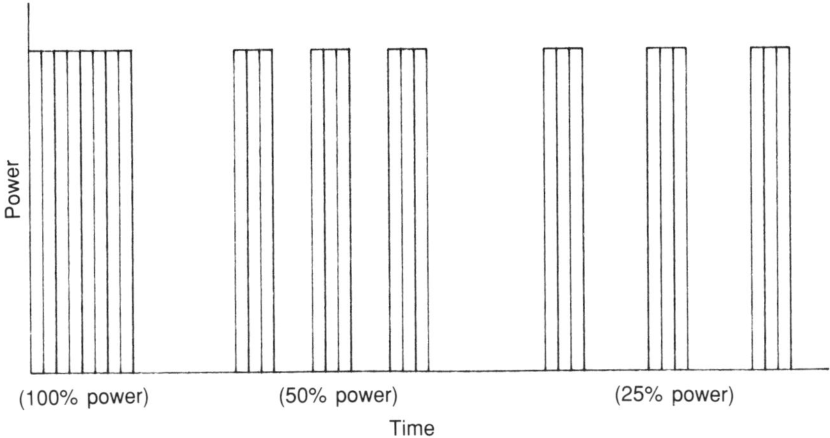

Fig. 6.9 Pulsing of RF-excited CO_2 laser.

but not for engraving, where frequencies as high as 10 kHz or more are needed.

Pulsing RF CO_2 lasers

It is considerably more difficult to pulse RF CO_2 lasers than HF or DC types because the impedance matching network which is necessary to ensure that power is coupled reasonably efficiently into the RF discharge will match effectively over only a narrow range of plasma tube currents. Consequently an RF laser is almost always pulsed from zero to maximum tube current. Power cannot be changed simply by changing the cw level of the tube current as in a DC laser. In the RF case, cw power is varied by pulsing the plasma tube current at a few tens of kHz and varying the mark-to-space ratio of groups of these pulses (Fig. 6.9).

This method of pulsing is relatively inflexible since the peak power of each pulse is always equal to the maximum cw power of which the laser is capable. The average power cannot be ramped up and down as is done with DC lasers so that, for example, circular welds and engraving cannot easily be carried out. In general the frequencies obtainable will be limited by the physics of the discharges and maximum pulse frequencies are generally similar to those obtained from DC lasers.

6.2.4 Stability

Stability is probably the most important feature of any laser, since if it is inadequate, applications are not reproducible. For most applications the stability of three different parameters is important. These are:

Power; Beam pointing direction; Mode.

All depend greatly on the thermal and mechanical stability of the resonator support structure.

Power stability

Power stability is important long term because lack of stability will adversely effect the repeatability of processing conditions. For practical purposes a power variation of ±5% over the unit working period is usually sufficient for most, though not all, laser applications. Many industrial lasers are more stable than this and most manufacturers of CO_2 lasers now quote a figure of ±2% for long-term stability. Generally the long-term stability of slow gas flow, fast axial flow and transverse flow lasers is comparable.

In the author's experience, however, there is a tendency for sealed lasers to be less stable, especially those at the upper end of their power range. This may well have two causes: there is a general tendency for sealed lasers to show a downward trend in power output caused by very slow contamination of the gas mixture and there is a tendency for higher-power sealed lasers to be undercooled because they are designed to be ultra compact. It is extremely difficult to design a laser which is both very small and very stable.

Pointing stability

Pointing stability is the directional stability of the centre of the laser beam. It is particularly important in applications involving the use of long beam paths, as are used for example, on large moving-optics cutting machines. Lack of directional stability is caused usually by mechanical or thermal movement of the structure which supports the resonator mirrors and is often accompanied by lack of power and mode stability, which generally share the same cause. Good-quality CO_2 lasers should exhibit a thermal stability of 0.5 mrad or better. This means that the beam centre will stay within a circle of 5 mm diameter at a distance of 10 m from the output window of the laser.

Mode stability

The transverse mode of a laser is, in effect, the distribution of energy in the output beam. Together with the divergence angle this is an important parameter effecting the processing ability of any laser. In general, good mode stability is more difficult to achieve than good power or pointing stability and usually has more effect on the laser's ability to carry out applications. When a mode is unstable it will generally change shape and become less round and symmetrical, usually causing the processing

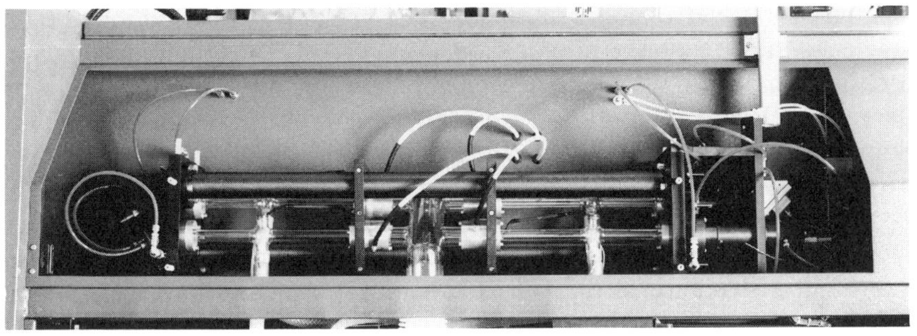

Fig. 6.10 Electrox fast axial flow CO_2 laser resonator structure.

performance to vary with the direction of travel of the laser beam or workpiece. The general energy distribution may also change so that the intensity is less in the centre of the beam and will tend to reduce the processing speed.

Designing for stability

The heart of any laser is the resonator and its supporting structure. The resonator includes all the resonator optics and, in the CO_2 laser, the plasma tubes which contain the discharge. The main purpose of the resonator support structure is to maintain the optics in the best possible alignment with one another under all operating conditions. It must also maintain the relative position of the plasma tubes with respect to each other and with respect to the optics, in all except transverse-flow CO_2 lasers. If these alignments are not maintained to the required degree then the laser will exhibit instability in the power output, in the mode, and in the beam-pointing direction.

At Electrox, a very large amount of effort has gone into designing and developing a stable resonator structure which has been used in various versions in all Electrox CO_2 lasers. In fact the Electrox type of resonator structure has almost become the industry standard for fast axial-flow CO_2 lasers.

The resonator support structure consists usually of two flat plates which carry the resonator optics and are often held apart by spacer bars or tubes which are made from a low-expansion material to minimize any length changes resulting from changes in temperature during operation. In spite of their being made of low-expansion material, the spacers are often contained in water-cooled jackets to further stabilize their temperature and to isolate them from small temperature changes inside the laser enclosure. A typical fast axial-flow resonator structure is shown in Fig. 6.10.

In fact, for most industrial applications of CO_2 only the stability of the spatial or transverse modes of the laser, i.e. the energy distribution at right angles to the direction of propagation, is important. The longitudinal mode, or the number of wavelengths of light between the resonator optics is largely irrelevant. Therefore it is not the length of the resonator which is important, but rather changes in the length of one resonator spacer relative to the others, which causes the optics to change their relative angles.

The resonator structure must also be isolated from mechanical forces which may be caused by dimensional changes occurring in the general mechanical structure of the laser or by vibrations from fans, pumps and even transformers. Again, the purpose of this is to prevent any twisting of one resonator mirror with respect to another. This is achieved usually by using some form of kinematic mounting. In fast axial-flow lasers the resonator structure is retained usually rigidly in the laser frame at one end only, and mounted at the other end by means of a single-point fixing which leaves that end free to rotate with respect to the laser framework. If the resonator support structure is connected to this point via a suitable spring plate then it is impossible for dimensional changes which occur in the laser frame during operation to transmit mechanical forces to the support structure and distort it. A spring plate is more effective than a bearing for this purpose because it does not exhibit 'stiction' and is also simpler.

Mounting of the optics on the resonator plates, and of cooling them, is very important in the search for stability. Ideally the optics should have their surfaces in contact with optically matching surfaces which are mounted permanently on the resonator support structure and never removed for cleaning or maintenance. All components supporting the optics on the resonator plates should ideally be made of high-conductivity metal so that the optics can be cooled as effectively as possible.

Of all the resonator optics the output window is the most critical, because it must transmit the laser radiation as well as reflect it. The slightest distortion of the shape of the window, whether by mechanical forces due to mounting or by thermally induced ones caused by dirt on the window or poor optical materials or coatings, will cause a corresponding distortion in the shape of the beam.

Ideally, the resonator structure of a CO_2 laser will be made completely from materials with the lowest expansion coefficient available. It will be mounted in such a way that it is completely isolated from all temperature changes and stresses of any kind, and it will be furnished with perfectly figured optics with perfect reflecting and transmitting characteristics. In the real world such a laser is impossibly costly to build and many compromises are of course necessary. The real job of the laser designer, especially when designing the resonator, is to see that he arrives at the best possible compromise in balancing the often conflicting demands of performance, cost, ease of servicing, user friendliness and many other lesser factors.

6.2.5 Building costs

At Electrox it is considered that the cost of building a range of CO_2 lasers depends strongly on the total number of parts used in each laser and on the number of different parts employed. It is therefore important to keep both of these numbers as small as possible. The company also believes that it is no more costly to design a component well than to design it badly. Thus the best design is usually the simplest one using the smallest number of well-designed (and well-made) components which will allow it to meet the specification required. To view the argument from the other side, the smaller the number of components which can be made to produce a satisfactory laser, the more can be spent on the design and manufacture of each component to make certain that it does its job well.

It is very important to consider the cost of building a range of lasers rather than a single model because customers use different sizes, power levels and types of industrial CO_2 laser depending on the application or range of applications which is being carried out. For example, Electrox builds a range of industrial CO_2 lasers with power outputs between 80 and 2000 W which will tackle a wide range of applications from perforating cigarette filter paper and drilling small holes in babies bottle teats to engraving rubber printing cylinders and cutting 15 mm thick steel plate. The range of lasers has been designed to use the smallest possible total number of separate parts. For example, the two smallest lasers, of 80 and 140 W power output are identical in their mechanical construction and merely use different length plasma tubes and power supply capacities. The 500, 800 and 1000 W lasers have the vast majority of components in common. The plasma tubes, gas recirculating blowers and high-voltage transformers are the only major components which differ from model to model.

6.2.6 Running costs

When considering the running costs of a CO_2 laser the most important thing is to remember to consider all of them, and not just those that may seem important at first sight. For example, there is a tendency to include gas costs, because gases are bought regularly and quite frequently, but to neglect the cost of replacement optics, which may be bought once or twice a year. The hourly cost of optics however, may easily exceed that of gases for some lasers. A list of running costs for a CO_2 laser should include at least the costs of the following items:

1. electrical power for laser;
2. electrical power for water-cooling system;
3. replacement optics;
4. laser gases;
5. processing gases if used;

6. other consumables such as pump oils, vacuum seals, electrodes and glassware;
7. servicing and maintenance.

6.2.7 Reliability

Today the CO_2 laser is regarded as just another machine tool by the majority of its users and, because of its high capital cost, reliability is not just expected, but demanded of it. The designer must keep this in mind at all times. Reliability is particularly important because CO_2 lasers are used normally in conjunction with complex and costly workhandling systems. The total cost of laser and system is usually over £100,000 and in some cases may approach £500,000 and when such a system is out of action a whole factory may be made idle. Thus the quality of parts, components and subsystems is of great importance to a designer of lasers and all of them must be used well within their specifications and intended lifetimes. Of equal importance are the test and quality control systems used by the laser manufacturer; these must ensure that there is little or no chance of a faulty part being discovered after, rather than before, delivery. In some cases the periods for which lasers are tested in the manufacturers works is as long as the time needed to assemble them.

6.2.8 Ease of servicing

Since a laser may be part of a £500,000 processing or manufacturing system, it is vitally important that when it goes wrong it can be made operational again as soon as possible. At all stages, the designer of CO_2 lasers must consider what will happen if a part or assembly should fail. From the designer's viewpoint, ease of servicing has three main aspects. The first is to decide what problems are likely to occur, the second, how to diagnose them and the third, how to put them right as fast as possible.

Ideally a designer should know the lifetime of every component in the laser. He can then include regular repair or replacement in the service and maintenance schedules and non-scheduled downtime will be unnecessary. In real life things are somewhat different. Some components have variable lifetimes even when used under apparently standard operating conditions; some undergo very gradual deterioration so that the effective lifetime is difficult to determine. Some customers have very different operating conditions from others and some customers carry out maintenance programmes much better than others. Some lasers are less than perfectly designed.

For these and other reasons all lasers fail unpredictably at some time so, if possible, the diagnostic system should indicate the likelihood of a failure well before it happens. In some cases this is relatively simple. For example, the viscosity of oil in a pump may be easily monitored and a signal given to carry out an oil change well before the viscosity has reached an unsafe level. Alternatively the oil can be changed at regular intervals. On the other hand, a bearing in a gas pump or fan may fail with very little warning and since changing it may be a lengthy job it is unlikely that laser manufacturers will recommend its replacement at regular intervals. Thus, design for ease of servicing is a complex problem.

The laser designer must first consider what sensors and transducers are to be fitted to the laser. This involves identifying expensive, delicate or inaccessible parts or assemblies which may fail or may be damaged by the failure of other components. In a CO_2 laser the parts in these categories include the output window, the blower in a fast axial-flow laser, the fan (or blower) in a transverse-flow laser and the high-voltage supply in any laser. In addition, failure of the main cooling system may damage almost any part of any laser. As well as diagnosing failure of important parts, the diagnostic system should indicate less-important problems which, although minor, can stop the system from operating. Thus many CO_2 lasers are fitted with sensors to show when a gas cylinder is empty or when a door has been left open.

Given that a choice of sensors has been made it is necessary to decide how to present information from the sensors to the operator. The method chosen at Electrox is to regard every possible occurrence which could prevent the laser from operating normally as a fault and to indicate it by the illumination of a red warning light. When the laser is operating normally and all red indicators are extinguished, a large amber light tells the operator that the laser is ready for use. Thus provided that there are no problems, the operator need only look for the presence of the amber light and can ignore the diagnostic system unless a fault occurs. The system uses only LEDs and simple logic integrated circuits. There is not a micro-processor or computer in sight. It is however much simpler and less ambiguous than many complex, computer-controlled systems in that every indicator light has a legend printed opposite which precisely describes the fault. For example, 'Shutter open', or 'Cooling airflow low'.

In some complex micro-processor and computer-controlled diagnostic systems the fault is simply diagnosed as 'a fault' and the operator is requested to repeat the start-up procedure. This is not to say that computer-based diagnostics systems are in any way inferior; in fact they are likely to be used in future on all lasers because of their great flexibility and simply because they can readily be integrated into a standard computer-based control system. However, it must be born in mind that any diagnostic system is only as good as the quality of information presented to the operator.

6.2.9 The control system

The control system of a CO_2 laser may consist of anything from a simple set of switches which operate contactors and relays in a sequence determined by the operator, to a complex computer-controlled system which automatically starts, runs, monitors and closes down the laser after initiation by the operator.

Any type of control system must be capable of evacuating the laser, filling it to the required running pressure, switching on the high voltage, opening the safety shutter to allow the beam out into the working area, and then shutting the system down safely and backfilling the vacuum system either with clean dry air or nitrogen, so that it may be re-started later with a minimum of delay.

Depending on the type of CO_2 laser and the degree of sophistication of both laser and control, the control system may also have several other functions. It may be used to run-up the laser automatically to the point at which the high voltage is switched on. It may be used to control valves which actively maintain the running pressure of the laser gas at a constant level, or to maintain mirror alignment automatically by operating motorized mirror adjusters in response to an error signal from a device which monitors the laser's beam shape. It may also, via a micro-processor, monitor several running parameters and produce an appropriate correction or warning signal should they depart from pre-determined ranges of values. The control system in many lasers will also be used to set the power level required for each application and to pulse the laser, sometimes via a separate pulse generator, if the laser is pulsable.

When, as in many applications, the laser is combined with a computer-controlled workhandling system, this computer may also be used to control the laser's control system. As, in most cases, both computers have sufficient power to control both the laser and the workhandling system, it is more cost effective to have one computer only. In some cases manufacturers of workhandling equipment are now specifying lasers without the control system and using their own computer to control the complete turnkey system.

6.3 INTEGRATION OF CO_2 LASERS WITH WORKHANDLING SYSTEMS

The CO_2 laser is generally regarded as the industrial workhorse of the laser world because it is used to carry out such a wide range of applications. For this reason CO_2 lasers have been integrated with a very large number of different beam and workhandling systems. In some cases integration is relatively simple. For example, if a CO_2 laser is used with a moving-laser or moving-workpiece workhandling system, it can be fitted with a fixed-beam focusing and gas jet system. The laser and head are simply positioned over the worktable in the correct position. The laser and table are then interfaced and the system should be ready to work. However, even in such simple systems it is very important to ensure that laser and workhandling machine are compatible, both electrically and electronically.

When a CO_2 laser is interfaced with a moving-mirror beamhandling system the integration can be more complex and problematical. Moving-mirror systems can be of the flat bed type used for cutting, gantry systems as used for car-body welding, robots, or articulated arm systems used in the medical field.

The first step in integration of a laser with a moving-beam system is to determine the clear aperture required for each of the optics in the system. For beam-handling mirrors a good rule of thumb is to measure the beam diameter at the mirror furthest from the laser, add an appropriate allowance if necessary for the pointing inaccuracy of the laser and then add a safety factor of about 50% to the resultant diameter to give the required clear aperture for each of the beamhandling mirrors.

It is a common mistake in specifying moving-optics systems to use mirrors which are too small. This results in the system being very difficult to operate, since any tiny change in beam diameter or pointing direction will result in the beam touching the edge of a mirror aperture. The resulting heating and distortion of the mirror mount, and diffraction of the reflected beam will then cause a change in the processing characteristic of the laser, which appears to be unstable.

A similar problem, often more severe, occurs when the clear aperture of the focusing lens is too small and the beam touches and heats the lens mount. In this case, in addition to problems of beam distortion, the result could also be a broken lens. A further problem with moving-optics systems is that the focused beam diameter may be different in different parts of the working area. This occurs because as the working beam length (distance from laser to focusing lens) increases, the beam diameter increases. As more of the available diameter of the focusing lens is filled by the beam, the lens aberration increases and the focused spot size increases.

Thus, usually, a moving-optics system will have a larger focused spot and often poorer processing performance, at that part of the working area corresponding to the longest beam path length. This effect can be minimized in two ways. A two-element focusing lens can be used which is aberration free and will eliminate the effect of changing beam diameter. Alternatively, a beam expander can be used to increase the initial diameter of the beam but reduce the divergence. On systems with long beam paths (say more than 6 m), a beam expander may well considerably reduce the difference in beam diameter from one end of the system to the other, so reducing the corresponding change in focused spot size. It will also reduce the largest beam diameter, and may allow the use of a smaller diameter (and lower cost) focusing lens. To obtain the smallest focused spot size over the whole table area it will be necessary to use both two-element lens and beam expander.

Such problems are very similar in both flat bed and gantry type moving optics systems. In robot and articulated arm systems similar problems occur and it is important to match the size of mirror and lens apertures to that of the beam. In addition however, problems may be caused by the number of mirrors employed. A typical articulated-arm system may have six mirrors in it so that losses can be significant unless dielectric-coated mirrors with reflectivities of 99% or more are specified. These may need a high reflectivity at visible wavelengths as well as at 10.6 μm if a He-Ne laser is used through the system as an alignment aid.

If careful steps are taken to solve the problems outlined above there is no reason why moving-optics laser systems should not be used as successfully as the simpler moving-workpiece ones. However, it should be borne in mind that a moving-optics CO_2 laser system is best specified where high speed and large workpieces are involved. It will not be the best choice where very high precision is needed.

6.4 OPERATION OF CO_2 LASERS: THE OPERATORS' POINT OF VIEW

In designing and engineering a new CO_2 laser, the designer will usually try to produce the highest possible specification in order to be better than the competition and also for reasons of personal pride. He will naturally be very concerned with such things as stability, mode shape and divergence, to say nothing of maximum pulse frequency, superpulse peak power and beam polarization.

Such parameters are naturally important but it should not be forgotten that a large number of industrial CO_2 lasers are used in jobbing shops which supply general industry with a large variety of mundane parts, from mild-steel washers to security locks. The owner or operator of such a shop is more likely to see performance in terms of 'how fast will it cut out 250 tab washers from 2 mm thick mild steel', than whether it is producing 100% TEM_{00} mode with 1.8% long-term power stability.

In other cases when a laser is bought by, for example, a large company and dedicated to a single application, the user is still interested first in the economics of carrying out that application and second in the detailed features of the laser, provided, of course that it will do his job with the quality and speed, and at the price which he needs.

For the owner and operator, the next most important features of a laser, after its practical performance, are probably its reliability and its ease of operation or user-friendliness. The ease and speed with which it can be maintained, serviced and repaired will also be seen as vital. The first time he buys a laser, the user may well be influenced considerably by the salesman's description of its high tech, high specification but the second and subsequent times he is more likely to want to know how often it needs to be serviced and repaired and who must do it. What are the running costs? How much helium does it use? How often must the optics be changed and how easy is it to change them?

To a company which has a CO_2 laser dedicated to one application, its flexibility, that is, the number of different jobs it can carry out and the speed with which the changeover can be made is irrelevant. However, to the average job-shop owner (actually I've not met a single job-shop owner who could be described as average) these things are of the utmost importance. To be flexible, a CO_2 laser must firstly have a large usable power range and this should be available instantly without any need to change working pressure and gas mixture, or without waiting for the resonator to stabilize after the power change. The bottom end of the power range should ideally extend down to a few watts so that the beam can be used directly for tracing the shape defined by an NC programme or for aligning the beam through a moving-optics beam handling system.

Furthermore the laser should be capable of changing instantly between cw operation, pulsed operation at any frequency, and superpulsed or extrapulsed operation without manual intervention by the operator. Only then can it take full advantage of the sophisticated control systems available today to work both automatically and flexibly at high production rates. There are also some very simple and low-cost ways of ensuring that a laser system can be both flexible and easy to use. For example, it is not difficult to ensure that cutting and welding gases can be changed over almost instantly. Neither is it costly to design a lens that can be exchanged in, say, a minute, for one of a different focal length.

Servicing and maintenance are other areas where a laser designer can score very heavily, both positively and negatively, with the users of laser systems. It is not very difficult to design the laser itself so that pump oils can be changed easily and so that resonator optics can be changed in half an hour instead of half a day. When actual repair work is involved the same principle applies. A laser manufacturer will score heavily with a customer if, when things do go wrong, any major assembly can be removed and replaced without the need to remove any other major assembly.

Finally, it is worth discussing the manufacturer's general approach to servicing and how it is effects the customer. It is possible for a manufacturer to insist that all servicing work is carried out only by his own well-trained service engineers, and that the laser is never touched by the customer except to carry out routine maintenance. This may be a good approach if the laser is almost completely reliable and the manufacture's service department is almost perfect. It is the approach adopted by manufacturers of white goods such as washing machines, and in such industries appears to be a successful one. In these industries this 'black box' approach works because the customer is usually non-technical and has little or no interest in the detailed workings of his or her washing machine or vacuum cleaner. In addition, it is a minor interest in their life as long as it is in working condition.

A laser-based machine on the other hand is owned and used by people working in a technical industry and in many cases it is playing a large role in the economics of the whole company. While it is true that the owner and operator are not really so interested in features of the laser such as its detailed mode structure, but more in its benefits to them – such as how fast it will weld a particular component – they can often gain great benefit from owning a machine which can largely be serviced and repaired by their own staff.

Since laser system manufacturers in general sell relatively small numbers of machines, in most cases a few tens each year, it is natural that service engineers must travel large distances to a customers' works, often to a different country. For this reason, if a laser manufacturer takes the 'black box' approach, his customers will often have to wait one or two days for an emergency service visit, often losing thousands of pounds worth of business in the process. If, on the other hand, he can carry out much of his own service and repair work, with telephone help from the manufacturer, he will often be working again, a few hours or even minutes after the breakdown.

This latter approach is not an easy one for the manufacturer to take, because for it to succeed he must put great efforts into making his laser 'user-repair friendly'. This is far from simple, since, even with training from the manufacturer, which is absolutely necessary, the customer's service staff will rarely be as expert as the laser system manufacturer's service engineers, who are carrying out the job full time. The laser company which opts for the route of encouraging customers to look after the laser machine themselves must work hard both to train them and to ensure that the laser can be serviced and repaired by someone who is not a laser repair expert. The author's view is that until lasers are perfectly reliable the do-it-yourself approach to service and repair is the better one.

6.5 CONCLUSIONS

The design of a CO_2 laser is above all else a matter of balancing and compromising. The laser must be technically capable of carrying out the required applications in competition with other available machines or processes. It must be competitive in price with other machines as well as other lasers. It must be safe to use, whether the operator is an expert or relatively unskilled, and it must be simple and quick to operate, service and repair. When it is considered that the CO_2 laser is a very subtle machine which in practice combines optics, large mechanical pumps, delicate mechanical components, high voltage and electronics, not to mention a great deal of piping and electrical wiring, it is a tribute to the many manufacturers of these devices that so many are used routinely to produce a wide range of parts and perform so many industrial processes.

7

CO_2 industrial laser systems and applications

C. Williams

7.1 ABSTRACT

The first practical demonstration of a laser device was in 1960, and in the following years the high-power carbon dioxide laser matured as an industrial machine tool.

Modern carbon dioxide gas lasers can be used for cutting, welding, heat treatment, drilling, scribing and marking. Since their invention over 25 years ago they are now becoming recognized as highly reliable devices capable of achieving huge savings in production costs in many situations.

This chapter introduces the basic laser processing techniques of cutting, welding and heat treatment as they apply to the most common engineering materials, reviews system and nozzle features, and describes a number of actual systems in some detail.

7.2 INTRODUCTION

The carbon dioxide laser emits a beam of infra-red energy that is nearly parallel, with a wavelength of around 10.6 μm. The diameter of this beam and its divergence varies from laser to laser, but typically, a 0.5 kW laser beam can be expected to have a diameter of 6–8 mm and a divergence of a few milliradians, while a 2 kW laser beam may have a diameter of 15–20 mm and a divergence of 1 mrad. Figure 7.1 shows an industrial laser of 600 W output power.

In over 25 years since its conception, the carbon dioxide laser has been used for a wide variety of techniques, some of which are highlighted in this chapter. Carbon dioxide lasers fall into four basic types, characterizing the laser gas flow, namely:

Laser Processing in Manufacturing. Edited by R.C. Crafer and P.J. Oakley.
Published in 1993 by Chapman & Hall, London. ISBN 0 412 41520 8

Fig. 7.1 Laser Ecosse MF 600P, 600 W industrial CO_2 laser.

1. Sealed gas;
2. Slow axial flow;
3. Fast axial flow;
4. Fast transverse flow.

In general, the sealed gas systems are used up to about 100 W. Above that, to about 2 kW, the slow axial flow has many advantages; it gives the most reliable, most cost-effective solution. From about 2 kW to about 8 kW fast gas flow must be used; usually fast axial flow is employed. This is, in general, a less attractive solution than slow flow due mainly to poorer reliability and higher relative cost. The blowers on fast axial flow lasers in particular, can present a problem if insufficient attention is paid to their specification and location. Unless carefully designed and mounted they can be unreliable, noisy, transmit high vibration to the system and have limited life. Above about 8 kW, fast traverse flow is most efficient, and powers in excess of 25 kW can be obtained with such a design.

7.3 LASER CUTTING

Laser cutting is the process where, due to the heat generated by the laser beam, the material is melted or vaporized. The molten and/or gaseous material is blown away by the assist gas which is usually coaxial with the

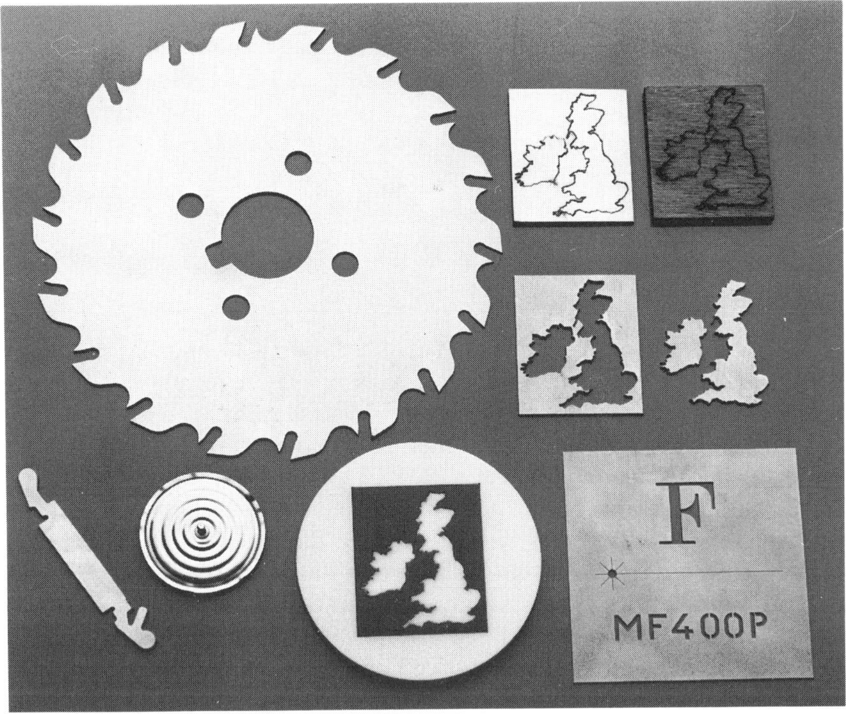

Fig. 7.2 Samples of laser-cut materials.

laser beam; in some cases the assist gas performs other functions (the role played by the assist gas is discussed in more detail in section 7.7). Despite the fact that the focused laser beam diverges rapidly after focus, thick materials can be cut and the resulting slot (or kerf) width can be surprisingly parallel. The width of the kerf depends upon the material composition, its thickness and other parameters associated with the laser system, such as power, speed, gas composition and pressure, etc. Kerf widths of 0.5 mm or less can easily be obtained in thinner materials. Figure 7.2 illustrates some laser-cut materials.

7.3.1 Advantages

The primary advantages for laser cutting can be summarized easily:

1. Very localized heat input;
2. High speed (compared with other methods);
3. Narrow kerf (typically less than 0.5 mm in most metals);
4. No cutting force;
5. No tool wear;

6. Ease of automation;
7. Different materials and thicknesses can be cut without resetting expensive tools;
8. Quiet operation;
9. Elimination of subsequent machining in most cases.

Drilling can also be accomplished with the CO_2 laser, but can roughly be considered to be a sub-set of cutting.

7.3.2 Materials

Not all materials lend themselves to CO_2 laser cutting. Copper, brass, bronze and aluminum are difficult to cut due to their high reflectance at 10.6 μm and their thermal conduction properties. However, they can be cut if a set of relatively rigorous conditions is met. Some non-metals, like polycarbonate, PVC and Kevlar can be cut with a CO_2 laser but produce hazardous fumes during cutting and the edge finish is generally not as good as most other non-metals. Excluding these exceptions the vast majority of common industrial materials can be cut by a carbon dioxide laser.

For example, using 600 W of power, 1 mm thick type 304 stainless steel can be cut at speeds in excess of 5 m/min with a typical kerf width of 0.2 mm. With a change to machine speed and assist gas pressure, the same laser can cut mild steel 6 mm thick. The edges will be square and dross free. Using a 1.5 kW laser, stainless steel up to 10 mm thick may be cut at production speeds with good quality.

Cutting non-metals may produce very interesting side effects. Acrylic can be given a 'flame finished' edge with no further finishing necessary due to a combination of gas pressure and laser power. Certain types of glass can be cut very cleanly, as can cardboard, ABS plastic, plastic film and cloth. Wood especially has an attractively 'stained' edge.

More recently, the development of multi-kilowatt lasers with genuine TEM_{00} beam profiles has allowed greater thicknesses to be cut. When the laser beam profile is truly TEM_{00} then the depth of focus that can be achieved is considerably greater than for lower-quality beams. For example, a 3 kW TEM_{00} laser can produce excellent cut quality and speed in various types of high tensile steel up to 28 mm thick. However it is likely that the main use of multi-kilowatt TEM_{00} lasers will be for high-speed (30 m/min) cutting of thinner steels to replace the blanking process and associated tool changes.

7.4 LASER WELDING

Laser welding is basically a simple process, requiring no filler wires, fluxes, electrodes or ancillary equipment. The power density used for welding is

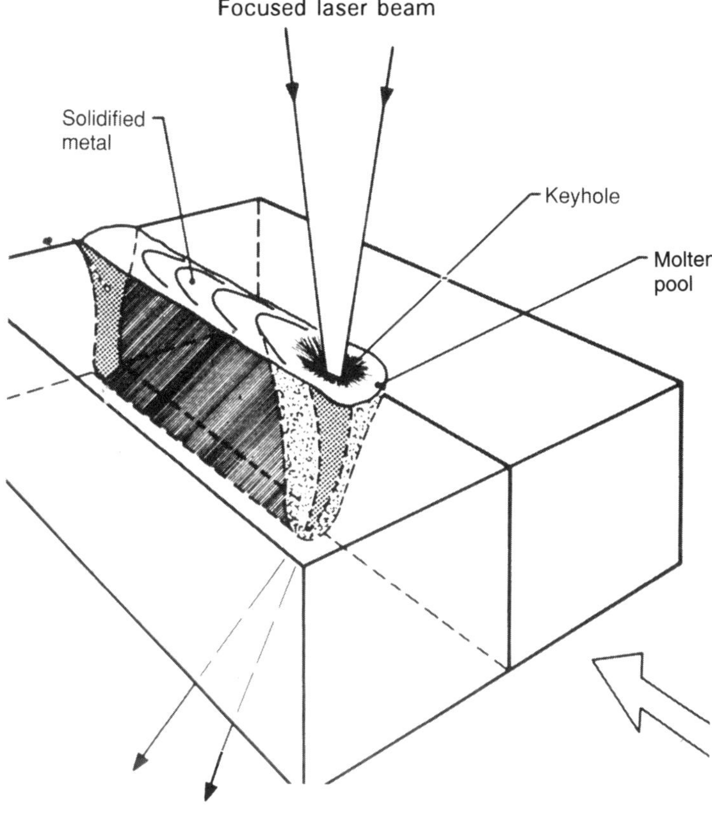

Fig. 7.3 Typical keyhole weld.

similar to that used for cutting, and so the same laser system may be used for both cutting and welding, and in some systems the conversion is software controlled. A keyhole welding process (Fig. 7.3), similar to electron beam welding, occurs with laser welding, however no vacuum chamber is required. If the parts to be welded are a close fit then no filler material is required. To prevent oxidation, argon or helium are used normally to shield the weld while it is being formed.

7.4.1 Advantages

The advantages of the laser technique, compared with arc or resistance welding may be summarized as follows:

1. Low heat input to the material, minimizing distortion;
2. Very rapid welds;

3. Easily automated process;
4. No weaving required;
5. The only consumable is shielding gas, apart from normal laser requirements;
6. It forms a narrow deep weld;
7. It can be applied in confined areas;
8. Access to only one side of the material is required.

7.4.2 Materials

The range of materials which may be welded by laser is vast and includes dissimilar alloys, carbon steels, plain steels, stainless steel, nickel alloys, copper and aluminium alloys. (These last two again require precise control over the laser and machine parameters to achieve good-quality results.) For a given material thickness, as a rough rule of thumb, laser welding requires five times the power that would be needed for laser cutting. Some important parameters required for cutting play no part in the welding process.

Laser welding can be carried out at high speeds; for example, stainless-steel washing machine drums at 6 m/min with a 1 kW laser, room thermostat assemblies at 9 m/min with a 400 W laser. However, for a maximum penetration, the speed must be reduced to less than 1 m/min. The plasma in the keyhole must not be allowed to escape the keyhole, and carefully designed gas jets enable it to be confined, otherwise too much power is absorbed by the plasma outside the material. Using such plasma control techniques, a 10 kW laser has welded 25 mm thick steel in a single pass. Laser welding can accommodate a wide variety of geometries, such as plain butt joints, overlap, tee-butt or tee-flared joints.

Welding non-metals can also be a very successful operation. Polyethylene, PVC, glass and man-made fabrics can be welded together. Very low powers of laser can be used for plastics; however utilizing a high-power laser leads to increased productivity. For example, one plastic bag manufacturer is using a 400 W laser to weld polythene bags at several metres per second.

Good fit up and jigging of the parts to be welded is essential. The gap in a clinch weld for example must be no more than 10% of the thickness of the material, otherwise an unsatisfactory weld will result. Often the advantage for laser welding is the fact that the penetration can be controlled. For example, in the welding of tube to plate parts for heat exchangers, the tube must not be perforated, so the weld penetration is only half way through the tube wall. On a car door clinch, the weld penetrates through two layers of metal and only half way through the outer skin, leaving no blemish on the outer skin. Figure 7.4 shows a similar weld displaying almost complete penetration.

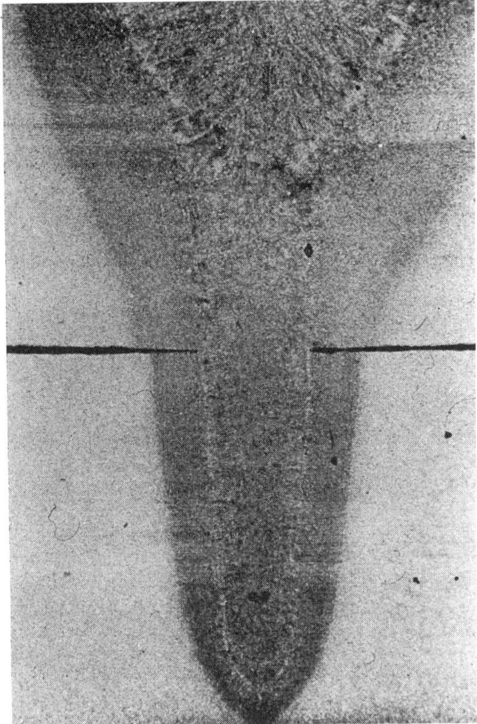

Fig. 7.4 Laser-welded sample.

7.5 LASER HEAT TREATMENT

Surface transformation using lasers covers many areas; for example, glazing, hardening, alloying and annealing. The three main areas are hardening, alloying and cladding.

7.5.1 The hardening process

Transformation hardening, the most commonly applied laser heat treatment, is induced by heating a metal above a critical temperature and then rapidly quenching it rather than allowing equilibrium phases to form by slow cooling. The most common transformations of interest are the formation of very hard martensitic structures in steel and the modification of cast iron. It is essential that surface melting does not occur and that rapid quenching does occur; the latter is achieved usually by conduction into the body of the material, i.e. it is self-quenching.

Transformation hardening is achieved by moving an unfocused laser

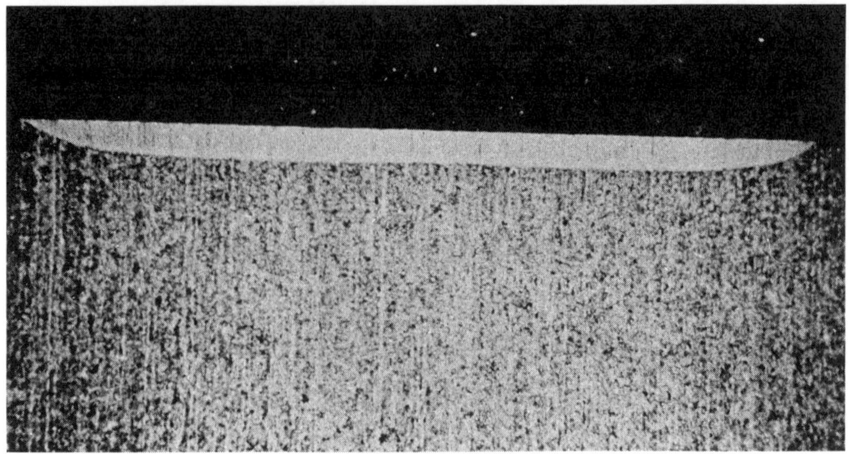

Fig. 7.5 Laser-hardened sample.

beam across the surface of the material, ensuring, by controlling power density and speed of traverse, that melting does not occur. In this way a thin surface layer is heated and then rapidly cooled by conduction once the beam has passed. Other scanning methods can be used to heat small areas, and fairly complex optical systems can by designed to move a partially focused beam over a wider surface area. Higher-order mode, or 'top hat', beam profiles can be used effectively here.

All steel that can be hardened by conventional methods can be hardened by the laser. The carbon content of the steel plays a great part in the final level of hardness achieved, but figures of Hv = 700 have been exceeded on 0.75% carbon steels. Cast iron can also be hardened, but much depends upon the distribution of the carbon in this material. If the carbon distribution is not uniform then the hardening process may have to be lengthened to try to allow for the carbon to diffuse throughout the material. Obviously cast iron of this type will require more stringent processing parameters.

Typical depths of hardening for steel and cast irons are 0.75 to 1.0 mm (Fig. 7.5), and the rate of hardening depends on laser power, material composition, absorbing layer on the surface, hardness depth required, etc.

7.5.2 Advantages

The advantage of using a laser to perform transformation hardening are:

1. Very localized heat input;
2. Negligible distortion of the component;
3. Treatment of specific areas as required;

4. Access to confined spaces is possible, e.g. hardening the inside of small bores;
5. Short cycle time required.

The obvious uses for such laser hardening systems are in the automobile or aerospace industry, for engine components, gears, transmission joints, etc., in the production of gun barrels, to harden the interior rifling for example, and general industrial tools and dies. One advantage of using a laser system is that it can first cut the shape required for, say, a trim die, and then go back over the top edge using a defocused beam to harden that edge. This has definite advantages in the production of tools and dies.

7.5.3 Surface alloying and cladding processes

Alloying is a process by which the surface of a material is first melted, then additional elements are added to this melted area so changing the composition of the surface. Cladding involves bonding a new material to the existing surface with minimal mixing of the materials.

Both alloying and cladding require higher power densities than hardening, usually by a factor 2 to 3; however similar optical systems can be used to perform both processes. Additional equipment is required to introduce the alloying elements to the surface; these elements may be powder, wire or sheet and can be positioned previously on the surface or be applied during the process.

7.5.4 Alloying

In surface alloying, the depth of the alloyed layer is equivalent to the depth of melting and is controlled by the power density and interaction time. Fine homogeneous structures are generally found due to the high cooling rates achievable with the laser process. Most studies have concentrated on alloying of chromium, carbon, niobium, vanadium or manganese to low-carbon and low-alloy steels, although work has been performed on treating aluminium alloys with silicon.

Advantages

The advantages of laser alloying are:

1. Low component distortion due to low heat input;
2. Treatment of complex shapes possible;
3. Fine control of melt-zone depth possible;
4. Accurate surface profile possible to save expensive alloying material and to minimize machining operations.

7.5.5 Cladding

While alloying relies on the complete mixing of the coating and substrate, cladding on the other hand requires minimum mixture of coating and substrate. Obviously there must be some mixing at the interface between the two materials for good adhesion, but with laser processing this can be reduced to about 2% (substrate into coating). Other methods of cladding, for example the submerged-arc process, may result in 50% mixing.

Common substrates include low-alloy steel and stainless steels, while coating materials have included Stellite, cobalt and nickel. Cladding thicknesses, using multi-kilowatt lasers, can be between 0.1 and 5.0 mm, achieved in a single pass, although thicker depositions may be achieved with multiple passes.

Advantages

The advantages of using a laser for cladding are:

1. Minimum dilution of the coating material;
2. Minimum distortion due to low heat input;
3. Accurate profiling of clad surface possible, minimizing waste of expensive cladding material;
4. Reduced machining due to accurate deposition.

7.6 SYSTEM TYPES

The laser is just an interesting laboratory component without an appropriate system around it. Each component in the system must be chosen for compatibility if a successful, reliable unit is to be achieved. Laser systems can be divided into three generic types (Fig. 7.6a, b and c respectively):

1. Moving laser
2. Moving workpiece
3. Moving optics

The choice of which type to use depends mainly upon the laser and the material to be cut. Obviously the laser in a moving-laser system must be compact and rugged enough to survive and operate properly. Speeds in excess of 10 m/min are usually required and accelerations up to 1 *g* are necessary. Moving-laser machines are relatively inexpensive and can be designed to accommodate large sheets of material. The load on the machine movement is a constant quantity (the laser) and so can be taken into account fully at the design stage. However, they are fairly slow and dependent on the fact that the laser can be moved and are often (but not always) restricted to flat sheet materials.

Moving-workpiece machines are also relatively inexpensive and can

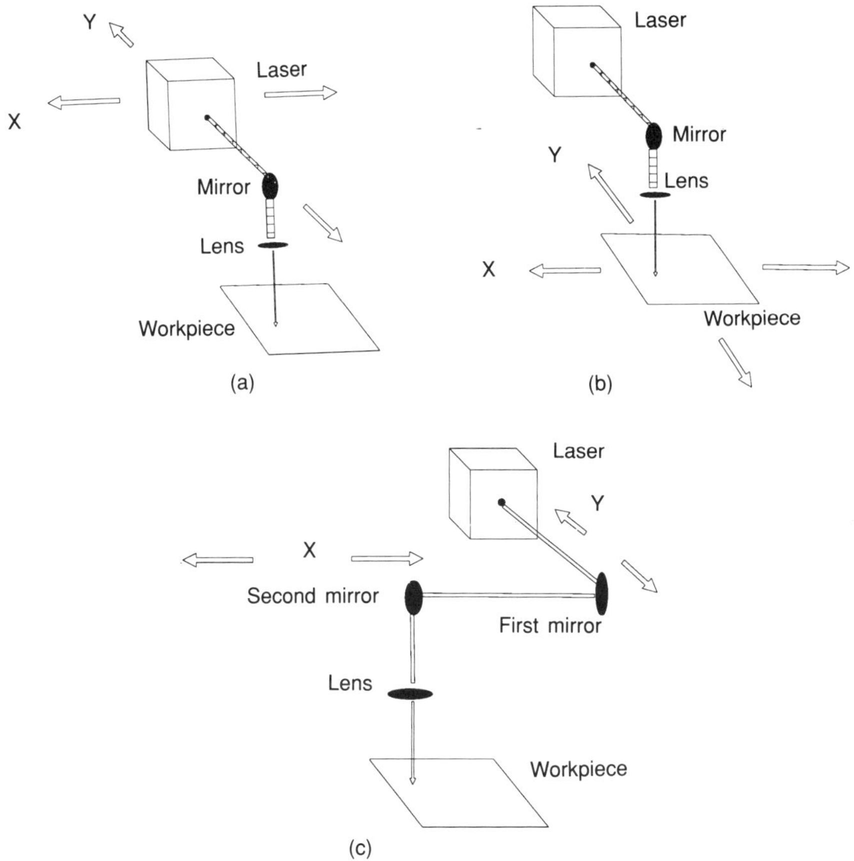

Fig. 7.6 The three generic CO_2 laser system types. (a) Moving laser, (b) Moving workpiece, (c) Moving optics.

accommodate a large heavy laser. They also can be made to operate quickly (up to 20 m/min). The main problem with them is that, while large sheets of material can be accommodated, even if they overhang the edge of the table, the weight of the material can easily exceed the capacity of the table. Also, the weight of the material to be processed varies, so the machine may not be operating within its inertial load specification. Three dimensional shapes can be accommodated by means of a 'wrist'-action nozzle together with a *Z*-axis slide.

Moving-optics machines are usually the most expensive but are capable of very quick movement (in excess of 100 m/min). A 'wrist'-action nozzle plus *Z*-axis slide converts a standard 2D machine to 3D, so that complex shapes can be processed. Drawbacks for this machine, apart from the price, are the facts that only certain sizes of sheet can be accommodated (if the sheet is only 1 mm over, then it is too wide), together with the problem of beam expansion down the optical path. As the beam propagates down a beam guide it expands with a set divergence. The mirror diameters must be calculated to be larger than the beam diameter at all times and if the divergence of the laser increases for whatever reason the mirror size could be too small, resulting in aperturing. The spot size in the focused beam is governed by, among other factors, the beam diameter incident upon the lens. If this diameter varies as the optics move back and forth, then the power density and hence quality or speed of process will vary. The system has to be designed carefully so that such variations are kept to a minimum.

Combinations of the three generic types exist and have been designed, in general, to reduce the problems associated with each individual type. Later in this chapter real systems will be discussed (section 7.9).

7.7 THE IMPORTANCE OF THE NOZZLE

7.7.1 Focusing and assist gas

To increase the power density of the beam sufficiently to achieve cutting, it must be focused to a small spot, typically 0.1–0.5 mm diameter. Focusing elements can be either transmissive or reflective (more simply, lenses or mirrors). Despite the high power densities present in such a small spot size, (many megawatts per cm^2), power alone is not enough to achieve good cutting, and an assist gas must be used to aid the cutting process; the role of this gas is described later in the chapter. In general, the higher the power density, the faster is the cut or the deeper the weld penetration, although other factors do influence these parameters to a greater or lesser extent.

A lens system is easiest to design and better suited to cutting, as the assist gas can be made coaxial with the beam rather simply. Above certain power densities, however, a lens system is unable to cope due to the power absorbed in the lens and a mirror focusing system has to be used. Mirrors are considered to be more rugged and resistant to damage than lenses. For example, mirrors can be conduction cooled readily via the back face whereas a lens must be convection cooled. Note that conduction rim cooling of lenses may make temperature gradients worse. However mirror nozzles with coaxial gas assist are difficult to design.

Many papers have been written about the effectiveness of nozzles, their shape, stand-off distances, pressures to be used in them, etc. Some of the data is sound and some controversial. However it is true that it is possible to have the best laser in the world but have bad results due to bad nozzle design or improper use. The lens nozzle system is simple in design, consisting of a focusing element, the lens, and a tip to offer up the focused beam and assist gas to the workpiece as shown in Fig. 7.7. The lens is

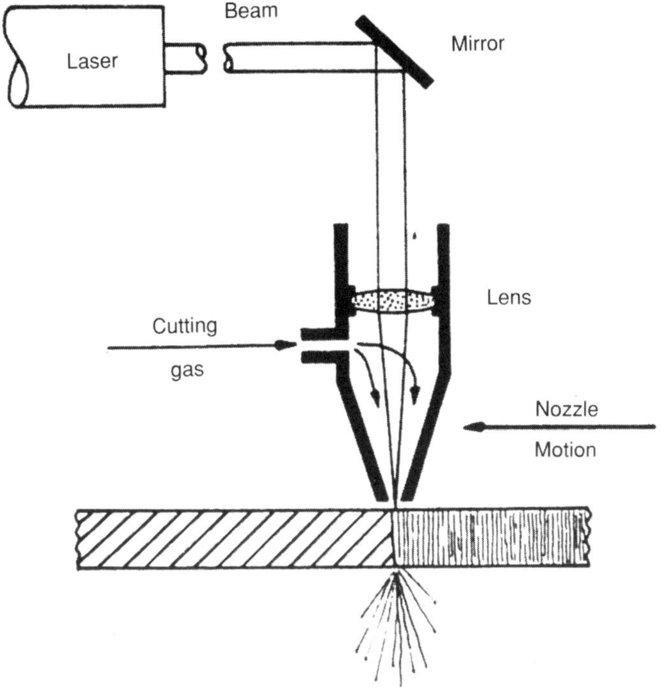

Fig. 7.7 Simple lens nozzle design.

usually kept at a constant distance from the workpiece by means of a height sensor. The latter can be either contact or non-contact depending upon the details of the actual system. In certain machines the height is controlled by a computer which 'knows' the geometry of the part and moves the *Z*-axis accordingly.

It is strange how often the theory behind a process is against the practical use of that process. This is true in the most simple way for nozzles. For cutting, the desired situation is to obtain the smallest spot size possible (giving the highest power density and hence fastest cutting speed) with the longest depth of focus (so that thick materials can be cut). For Gaussian beams the spot size is proportional to the lens focal length divided by the beam diameter on the lens, that is:

$$\text{Spot diameter is proportional to } \frac{\text{Lens focal length}}{\text{Beam diameter}}$$

whereas the equation for the depth of focus follows the form:

$$\text{Depth of focus is proportional to } \frac{(\text{Lens focal length})^2}{(\text{Beam diameter})^2}$$

As can be seen, the two equations work against each other. To obtain the smallest spot size and therefore have the highest power density, the focal length must be small. To obtain the greatest depth of focus the focal length must be large. So a constant compromise must be made to ensure that the correct cutting conditions are maintained.

Mirror focusing systems are of two main types, the *Z*-fold and the 'figure of 4'. These are used for welding or surface transformation processes, where high powers are required and coaxial gas jets are not necessary. Due to the higher powers used, mirror systems can afford to be longer focal length than lens systems, so, added to the fact that mirrors are generally more rugged than lenses in any case, this makes mirror systems much less prone to damage from splatter from the workpiece.

7.7.2 Assist gas

The assist gas has many uses:

1. It removes molten and vaporized material from the cutting zone.
2. It prevents molten material and vapour from contaminating the focusing lens.
3. It can be used to cool the focusing lens.

If the assist gas contains oxygen, i.e. oxygen itself or compressed air, then an exothermic reaction occurs when cutting ferrous metal and cutting speed is increased. When cutting non-metals, where burning is a problem, or to prevent oxide build up on the cut edge of ferrous components, an inert gas must be used.

When oxygen assist is used, the laser first heats the metal, the metal then burns rapidly advancing ahead of the laser beam, and the burning then ceases, only to be reinitiated as the beam catches up. This leads to characteristic striations on the cut edge, which is of course oxidized. When inert gas is used the metal is merely melted and the molten material blown clear by a high-pressure (20 bar) jet (the melt shear mechanism). This process is slower than oxygen-assisted cutting but often has a smoother cut edge and is oxide free; two pieces of stainless steel cut by this so-called clean-cut method may be laser welded back together easily.

The assist gas is usually arranged to be co-axial with the beam. This simplifies profile cutting and makes the cut edge more uniform.

A shield gas is used where oxygen in the atmosphere would have an adverse effect on the process, for example welding or cutting some materials, etc. In the latter case, the cutting of titanium requires the cutting point to be surrounded by an inert gas to prevent oxidization; the cutting of a combustible material also requires careful shielding by gas. For welding, a complicated series of gas jets can be employed to remove the effects of oxygen from the laser beam/material interaction point. Helium is the best gas for weld shielding, but it is expensive, so argon or nitrogen are often used. New gas mixtures are now being marketed, claiming to have nearly all of the benefits of helium but at lower cost. As has been said before, nozzle design is an art, with many different opinions on the correct way to design a nozzle for a particular application and laser. The reader is urged to take the above section as a guide only which just touches the subject.

7.8 OTHER SYSTEM CONSIDERATIONS

Pulsing

For cutting large shapes which are relatively simple in outline, a continuous wave (CW) laser can be used. Here the beam is continuous, on/off control being accomplished by means of a shutter.

In general, however, cutting involves producing intricate shapes and so a laser capable of pulsing is the best solution. The average power of the laser can be controlled by turning the laser on and off electronically thousands of times per second. In this way a precise amount of power can be incident upon the surface of the material resulting in perfect cutting every time. Without pulsing, too much power may be used, resulting in wide kerf widths, large heat-affected zones and even burn-out of the material. The duty cycle of the pulsed waveform can be controlled by a CNC which uses the speed of the table as input.

Mode and divergence

Laser beam quality is also to be considered when cutting. Two lasers of the same power may cut very differently depending upon the profile of the laser beam. The best profile for cutting has been determined to be of Gaussian form, in laser terminology called TEM_{00}, which is the lowest-order mode of operation of the laser and results in the best cutting performance. Higher-order modes, non-Gaussian in shape, result, in general, in non-optimum cutting either by virtue of lower speed, poorer quality or less penetration. Theoretical diametric cross-sections of the two low-order modes are shown in Fig. 7.8.

Polarization

The laser beam output can be randomly or linearly polarized, depending on laser design. For a linearly polarized beam the speed and quality of cut in metallic materials will vary depending upon the relative orientation between the direction of cutting and the polarization vector. To overcome this variability, optics must be introduced to polarize the laser beam circularly; in this way uniform metal cutting is achieved, regardless of

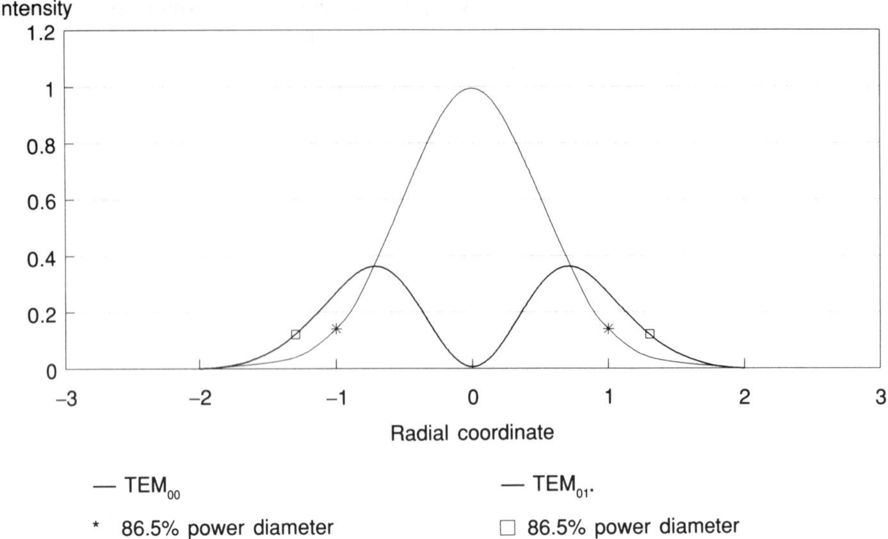

Fig. 7.8 Theoretical intensity profile for TEM_{00} and TEM_{01}* modes.

direction of cut. New research into welding with linearly polarized beams has shown that process speed can be increased if the direction of polarization and direction of weld are the same.

CNC

The CNC controlling a system must be capable of carrying out a stored program correctly and of moving the relevant component, be it laser, workpiece or optics, smoothly in the desired way. New generations of CNC are now tailored to laser-processing specifically, controlling assist-gas selection and pressure, laser power settings, pulse settings, focus adjustment, etc. They are also becoming more user friendly and are carrying more on-board memory. Off-line programming is common, with data being downloaded via direct link or floppy disc.

Table characteristics

Whether it is the table that moves or the laser or the optics, the movement must be smooth and be capable of high speeds and accelerations. Speeds of 20 m/min are now being required for profiling on moving-table systems; this means not only a rugged table with the material being lightly clamped, but also a fast processor to calculate the contour required. Speeds in excess

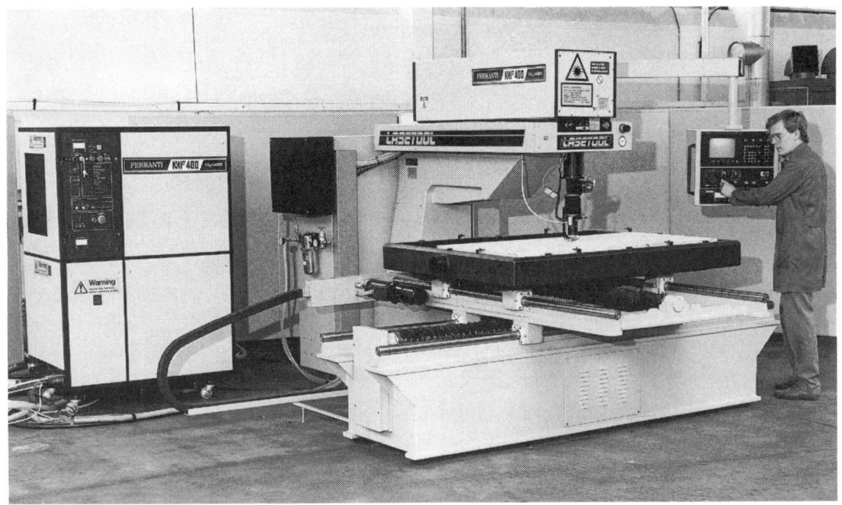

Fig. 7.9 *X-Y* table system with Ferranti (Laser Ecosse) 500 W laser.

of 100 m/min can be achieved with moving-optics systems, but the penalty is price.

High acceleration is a requirement so that the machine does not 'linger' in the corners of a profile too long. This can result in burn out of the corner and obviously is to be avoided. To take advantage of the inherent accuracy of the laser, the machine accuracy should be of the order of 50–100 μm for really precise work. However the machine accuracy requirement comes from the type of work required. Some requirements call for very tight tolerances, while others specify a few millimetres!

7.9 ACTUAL SYSTEMS

A few actual systems are now described, detailing the functions they perform.

7.9.1 Moving workpiece

Figure 7.9 shows a moving workpiece system based on a Ferranti (Laser Ecosse) 500 W CO_2 laser and a Wadkin *X-Y* table, which was used originally for routing. This type of system is simple to use and maintain, very versatile and ideal for job-shop applications where different tasks are performed during the day. It can be fitted with a small rotary axis for tube cutting or welding.

Figure 7.10 shows a 1 kW laser integrated into a very compact system. This was custom-designed and built by Ferranti (Laser Ecosse) for NEI

Fig. 7.10 *X-Y* table system with Ferranti (Laser Ecosse) 1 kW laser.

Parsons in Newcastle. Here the task was to de-flash drop-forged, stainless-steel turbine blades. The nozzle on this system is controlled by computer to move up and down under program control, i.e. there is no height sensor as such. The ability to maintain focus relies solely on the part being the same shape and in the same position as the computer expects it to be. The system has been working for a number of years on the shop floor at NEI.

7.9.2 Moving laser

Figure 7.11 shows a moving-laser system at Arden Dies in Stockport. Here the 500 W lasers are fitted to specialized, highly accurate die board cutting machines. The use of laser systems in the die board cutting industry has revolutionized this business, reducing the time required to make the dies by about an order of magnitude.

Figure 7.12 shows a 1 kW laser fitted to an Asquith machine installed at GEC in Manchester. Its usual task is the cutting of stainless-steel turbine blades; however it is shown here cutting a large saw blade. Not shown is a large rotary axis for cutting the circular rings into which the turbine blades are located.

Fig. 7.11 Die board cutting system with two Ferranti (Laser Ecosse) 500 W lasers.

Fig. 7.12 Large moving laser system with 1 kW laser.

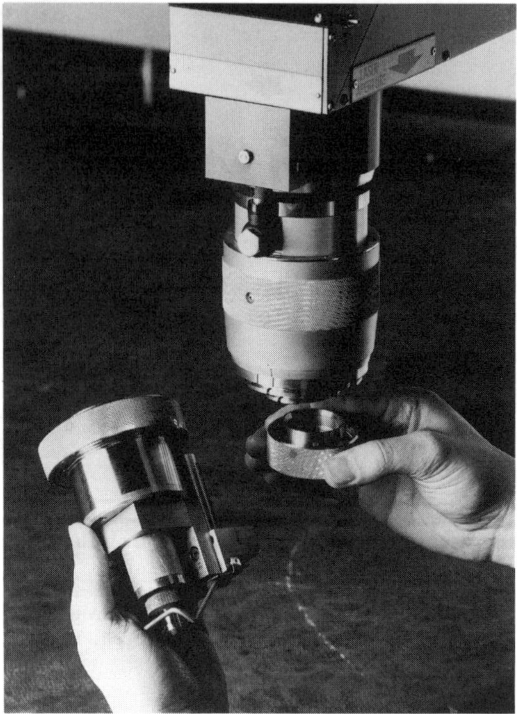

Fig. 7.13 Nozzle detail from Fig. 7.12.

Figure 7.13 shows a detail from this machine, the nozzle, designed for ruggedness and ease of maintenance. The lens is shown, in its holder, and is exchanged easily by removing the lower barrel. Positive-locking, bayonet fittings are used to ensure good fit and quick replacement. Focus can be adjusted manually to compensate for different lenses, and thereafter is maintained by the height-following system. The copper nozzle tip is precision made so that the mixing of the beam with the laser gas results in the gas being coaxial with the beam.

Figure 7.14 is really a combination of moving laser and moving optics, in that the complete beam delivery system moves into position as well as the laser. It shows a Ferranti (Laser Ecosse) 500 W laser integrated into a system designed and built by Robomatix of Israel installed into the Volkswagen Golf/Polo production line. The task is to cut left/right-hand-drive modifications. Cars for right-hand drive countries stop in the laser cell and get a hole cut under the dash and also the windscreen wiper holes cut (this latter task is performed by another laser not shown in the photograph).

Fig. 7.14 Ferranti (Laser Ecosse) 500 W laser on production line at Volkswagen.

7.10 CONCLUSIONS

More and more industries are installing laser systems for a wide range of applications. The majority of the applications involve cutting; however, as welding and surface transformation by laser are becoming established, they will grow in popularity.

Cutting requires good beam mode, circular polarization, stable mode, good pointing accuracy and stable divergence. Many types of system are available, each offering advantages for the end user.

Welding and heat treatment usually require higher powers, no polarization (although some welding and heat treatment developments are using linearly polarized beams), stable mode and divergence. Higher-order modes are more suited to these applications than TEM_{00}, although the Gaussian mode can be used.

Pulsing is essential for fine profile cutting in metals and has advantages in some welding applications.

The wide range of machines, matched with the appropriate laser, provides a very broad spectrum of industrial machine tools which can be selected to suit many applications.

No doubt the future will see the laser trend continuing and many more systems installed in all kinds of industrial areas.

8

Excimer lasers: principles of operation and equipment

M.C. Gower

8.1 PRINCIPLES OF EXCIMER LASERS

8.1.1 Background

The excimer laser was first operated in 1975 approximately 13 years after the invention of most other common types of lasers such as carbon dioxide (CO_2), neodynium yttrium aluminium garnet (Nd:YAG), ruby, diode, dye and He-Ne.

The term 'excimer' is short for 'excited dimer' where 'dimer' refers to a diatomic molecule such as N_2, O_2, H_2, and C_2 in which both of the atoms are the same. Since the two atoms are not the same for the type of diatomic molecular lasers we are concerned with in this chapter, they should strictly speaking be called 'exciplex' – short for 'excited complex'. However it is now common usage of the term 'excimer laser' to refer to the class of rare gas halide molecules such as argon fluoride (ArF), krypton fluoride (KrF), xenon fluoride (XeF), krypton chloride (KrCl) and xenon chloride (XeCl), that are the active laser species.

One learns in early school chemistry that rare gases are called 'inert' because they do not readily react with other atoms or molecules to form new molecules. It was not until the early 1970s that it was found that if a rare gas atom (He, Ne, Ar, Kr or Xe), which normally has all its permissible electron orbits around the nucleus filled with electrons, is electronically excited by changing the orbital motion of one or more of the electrons or by removing one altogether, then it can become highly reactive to other atoms or molecules in the vicinity. The excited states of rare gas atoms are in large orbits and see a core of unit net charge which makes them appear similar to their neighbouring one-electron alkali atom partners in the periodic table. Apart from the vacancy in the core, the configuration of

Laser Processing in Manufacturing. Edited by R.C. Crafer and P.J. Oakley.
Published in 1993 by Chapman & Hall, London. ISBN 0 412 41520 8

the electrons of the first excited states of He, Ne, Ar, Kr, and Xe correspond directly to the ground state configurations of Li, Na, K, Rb, and Cs respectively.

Since the chemical reactivity of atoms is determined principally by the configuration of their outermost electrons, the excited states of rare gas atoms can be as highly reactive as alkali atoms. In particular, ions of positively charged rare gases and negatively charged halogens (F, Cl, Br or I) that have an extra electron in their outer orbit, very strongly bind together to form an excited state. This 'Coulombic' ionic binding is responsible for holding together salt molecules such as sodium chloride (NaCl), potassium iodide (KI) and caesium bromide (CsBr) in their ground unexcited states. Similarly, excited rare gas halide molecules are formed by the strong mutual attraction of positive and negative ions of rare gas and halogen atoms.

Shown graphically in Fig. 8.1(a) and (b) is the binding energy for this attraction as the nuclei come closer together to form the rare gas halide molecules KrF and XeF. Although the potential energy well of the excited ionic 'B-state' is deep when the rare gas halide molecule is created, it is unstable in the sense that 5–15 nanoseconds after its formation an ultraviolet (uv) photon is emitted that puts the molecule into its ground 'X-state'. Since the unexcited atoms show little affinity and in some cases strongly repel one another, they can fly apart in a fraction of a picosecond (10^{-12}s). In such cases, the ground state of a rare gas halide molecule essentially does not exist. Because of its sparse population, such a repulsive or weakly bound state makes an ideal lower level for a laser transition. The difference in population between the upper and the lower states – the population inversion – which determines the amplification gain of a laser is then maximized. It is the B $\Rightarrow$ X transition of the molecule that is used to produce laser action in rare gas halide excimer lasers.

As well as being formed by the mutual attraction of positive rare gas and negative halogen ions, rare gas halide molecules can also be produced by excited rare gas atoms in a relatively long lived metastable state colliding with a halogen bearing gas molecule such as F_2 or HCl. In the collision a halogen atom can be plucked from the molecule by the excited rare gas atom. As shown by the potential energy curves in Fig. 8.1, the molecular states produced by these and other processes may cross one another and, just as in a game of bagatelle, the molecules can readily change back and forth between them as they try to drop by collisions into the one with the lowest potential energy – in this case the ionic state bucket. For some molecules such as the halides of He and Ne, curve crossing occurs with excited states or the ionization limit of the halogen atom near the bottom of the B-state bucket and it becomes leaky – once formed by ionic attraction, molecules curve cross and fly apart to form excited or positively ionized halogen atoms in processes called 'predissociation' and 'autoionization' respectively. For practical purposes halides of these atoms do not exist.

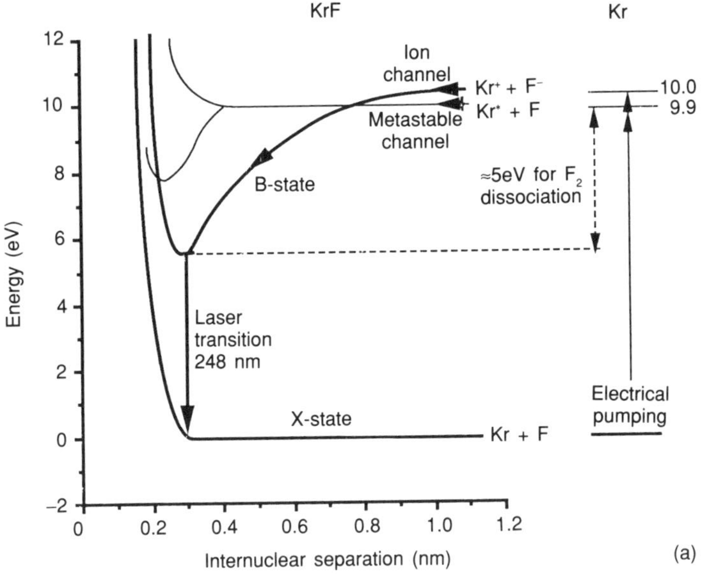

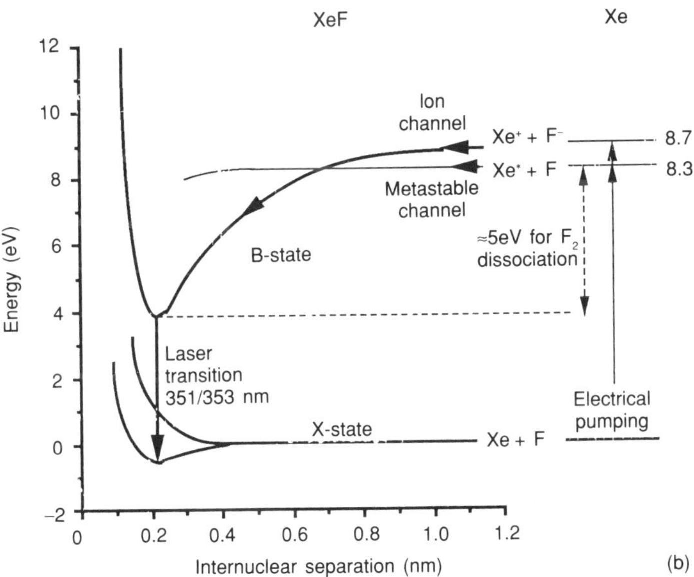

Fig. 8.1 Binding energy versus internuclear separation for the rare gas halide molecules (a) KrF and (b) XeF showing the B $\Rightarrow$ X laser transition and the excited atomic levels of the rare gases responsible for the formation of the B-state.

Table 8.1 Wavelengths in nm of B ⇒ X emission in rare gas halide molecules

	F	Cl	Br	I
Ne	*108P*	*A*	*A*	*A*
Ar	**193**	175P	*161P*	*A*
Kr	**248**	**222**	*203P*	*185P*
Xe	**351/353**	**308**	282P	*254P*

A and *P* indicate autoionized and predissociated B states, *italics* indicate no laser emission (only fluorescence) observed and **bold** the most important laser transitions.

The wavelengths of the most important rare gas halide molecules useful for laser purposes are shown in Table 8.1.

8.1.2 Emission spectrum

When an electronically excited molecule changes from one state to another by emitting a photon, the vibrational and rotational motions of its heavy nucleii tend to remain relatively unaffected by changes in the configurations of the much lighter orbiting electrons involved in the transition. Thus photon emissions and absorptions caused by rearrangements to the electron cloud tend to appear as vertical transitions on potential energy diagrams like Fig. 8.1. Since for rare gas halide molecules the ground state is repulsive or only weakly bound, the emission of the UV photon from the ionic B-state occurs over a relatively wide wavelength band. In Fig. 8.2(a) and (b) we show the spectrally resolved B ⇒ X laser emission from KrF and XeCl molecules in the wavelength regions of 248 and 308 nm respectively. The ground X-states of ArF, KrF and KrCl molecules are repulsive and so in this state cannot be characterized by vibrational or rotational motion of their nucleii. A broad featureless continuous laser emission over a ≈0.4 nm wavelength spread is produced. The several discrete bands observed on the KrF laser spectrum in Fig. 8.2(a) are due to absorption by impurity species that build up in the laser tube as the gas mixture ages and becomes stale. For the xenon halides, XeF, XeCl and XeBr the neutral unexcited atoms are weakly attracted by one another, leading to a weakly bound ground X-state in which the nucleii are free to vibrate and rotate. As can be seen in Fig. 8.2(b), for these molecules laser emission is produced at discrete wavelengths that correspond to the rotational and vibrational motions of the nucleii following the reorientation of the electrons after the photon is emitted.

8.1.3 Laser kinetics

The most important gas collision and photon processes involved in producing excited rare gas halide molecules suitable for laser action are summarized in Table 8.2.

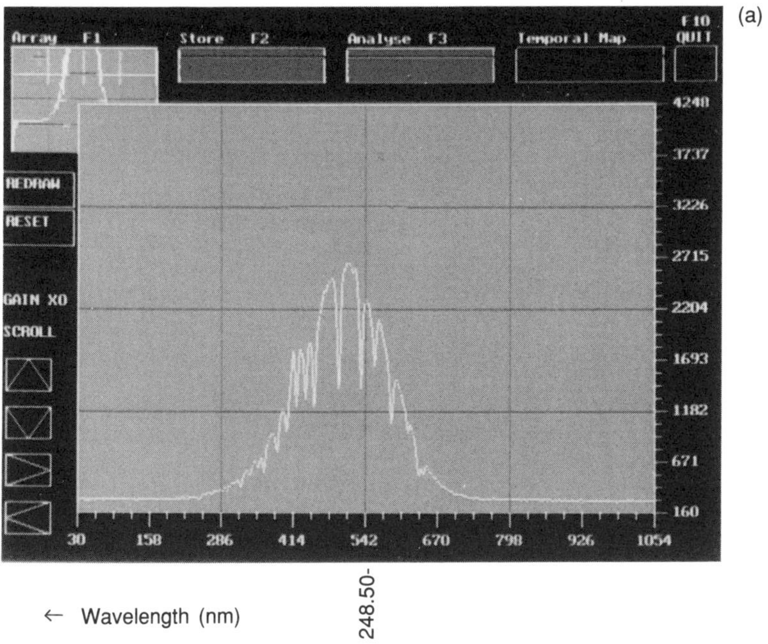

XeCl

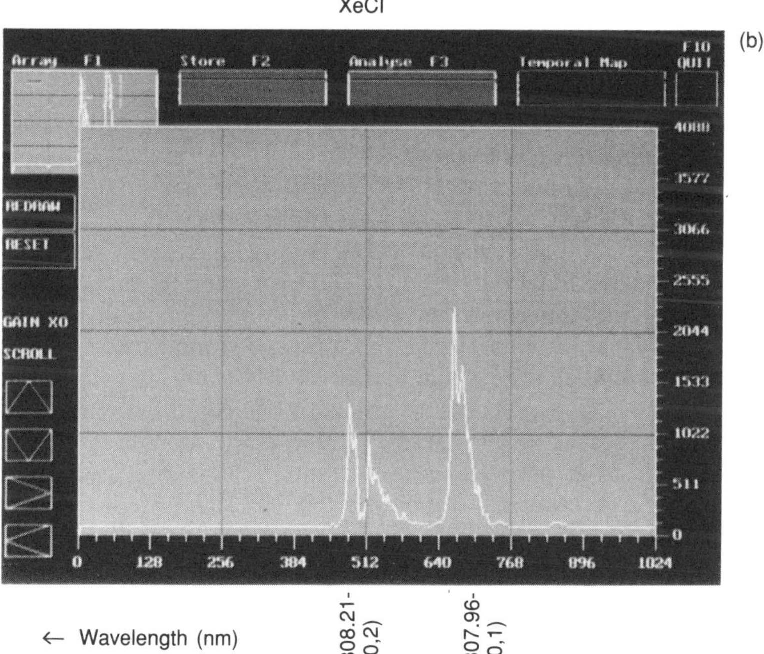

Fig. 8.2 B ⇒ X laser emission spectrum at (a) 248 nm due to bound–free transition in KrF, (b) 308 nm due to bound–bound transition in XeCl.

Table 8.2 The most important gas and photon kinetics processes of discharge-excited KrF lasers

	Process	Description
Pumping		
(i)	$e^- + Kr \nearrow Kr^+ + e^- + e^-$	Positive rare gas ion production
	$e^- + Kr \searrow Kr^* + e^-$	Rare gas metastable production
(ii)	$e^- + F_2 \Rightarrow F^- + F$	Negative halogen ion production
(iii)	$Kr^+ + F + M \Rightarrow KrF^* + M$	KrF production
(iv)	$Kr^* + F_2 \Rightarrow KrF^* + F$	KrF production
Stimulated Emission		
	$KrF^* + h\nu \Rightarrow Kr + F + 2h\nu$ (248 nm)	Laser emission
Losses		
(i)	$KrF^* \Rightarrow Kr + F + h\nu$ (248 nm)	Spontaneous emission
(ii)	$KrF^* + M \nearrow Kr + F + M$	Collisional deactivation
	$KrF^* + M \searrow + Kr \Rightarrow Kr_2F + M$	Collisional deactivation producing Kr_2F
(iii)	$X + h\nu$ (248 nm) $\Rightarrow X^*$	Laser photon absorption

M and X are third body collisional partners and impurity molecules respectively.

Although we have chosen Kr and F as the rare gas and halogen atoms, the kinetics for other excimers are similar and obtained by appropriate substitution of atoms. Electrons that are produced either in a high-voltage electric discharge or by secondary creation with an electron beam in an appropriate mixture of rare and halogen gases are used to collisionally excite the rare gas atoms to metastable electronic states or to form positive ions. Some of these electrons are also captured by the electronegative halogen-bearing molecules (F_2 or HCl) in the gas mixture. The KrF molecule in the excited upper laser B-state is then formed by the positive and negative ions combining together or by an excited Kr atom plucking a fluorine atom from the F_2 molecule. Because the ground state is repulsive no atoms are formed in the ground lower laser X-state. As with all lasers, if there is sufficient population inversion to overcome the optical cavity and medium losses within a suitable optical resonator then laser action will occur, in this case at a wavelength of 248 nm. Although it would require prohibitively large powers to pump the discharge and excessive gas and component cooling requirements, in principle excimer lasers can produce laser radiation continuously. All practical devices are pulsed and produce bursts of UV laser light each lasting 10–30 ns at frequencies as high as several thousand times a second.

By not adding the rare gas to the chamber and discharging in a mixture of fluorine and helium buffer only, molecular fluorine can also be made to produce laser action at the extremely short wavelength of 157 nm. Because,

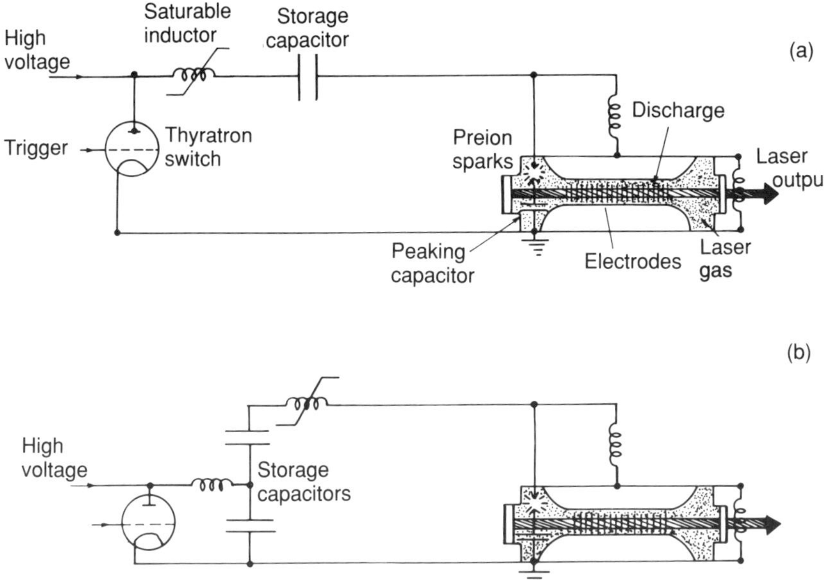

Fig. 8.3 Typical electrical circuits of UV preionized electrical discharge excimer lasers showing (a) matched and (b) L–C inverter circuits.

at wavelengths shorter than ≈190 nm, molecular oxygen is highly absorbing to light, air must be expelled from the beam path of this laser by either evacuating or flushing it with gases such as nitrogen, helium or argon. This makes this excimer laser rather inconvenient to use and beam delivery systems for applications tend to be cumbersome.

8.1.4 Devices

As shown in Fig. 8.3, to excite the fuel gas in the laser chamber most commercial excimer lasers use a 50–100 ns duration, 35–50 kV voltage pulse applied across electrodes that are orthogonal to the directions of the laser radiation and the gas circulation flow. The ensuing self-sustained electrical discharge in the gas passes peak current densities of up to ≈1 kA/cm^2. Storage capacitors are DC charged to the working voltage and then, with either matched or L-C inverter circuits, small doorknob ceramic capacitors situated close to the electrodes are pulse charged through a thyratron switch. The close proximity of the secondary capacitors to the electrodes provides a low-inductance pathway for current to flow through the gas and allows an extremely fast rising (≈1 ns rise-time) high-voltage pulse to be applied to the electrodes. The gas in-between initially has a high impedance and requires ≈15 kV/cm to break down to form a conducting plasma. Atoms in the gas become ionized in collisions with electrons to produce the high densities of ions and electrons necessary for conducting the current through what becomes a discharge with an impedance of ≈0.1Ω. Inductors with saturable magnetic cores are often incorporated into the circuit to make the transfer of energy more efficient.

In the matched circuit shown in Fig. 8.3(a) the impedance of the secondary discharge circuit is matched to that of the primary energy storage circuit so that maximum energy transfer occurs between the two with a minimum of aftercurrent ringing through the thyratron. The saturable inductor provides a 'magnetic assist' that slows the current rise and curtails potentially damaging aftercurrents from passing through the expensive thyratron. If the two storage capacitors shown in the *L-C* inversion circuit in Fig. 8.3(b) have the same value, then when the thyratron closes, nearly twice the charging voltage is applied to the electrodes, so DC charging voltages can be reduced. In this case the saturable inductor also acts as a 'magnetic switch' to compress the voltage pulse applied to the electrodes. Wall-plug efficiencies for the conversion of electrical to UV laser photon energy in such discharge-excited excimer lasers can be as high as 2–3%.

Because of the high affinity of halogen gases for electrons, avalanching of electrons by over 12 orders of magnitude from the background noise density created by random events such as cosmic ray ionization, to the required discharge densities of $\approx 10^{16}/cm^3$, cannot proceed efficiently and uniformly. Unless the discharge region is pre-seeded uniformly with electrons to densities of $\geq 10^8/cm^3$ prior to applying the main discharge voltage so as to reduce the amount of electron avalanching, strong arcing and non-uniform excitation throughout the volume will occur and prohibit laser action from taking place. Such electron pre-seeding is achieved usually by flooding the volume with UV radiation supplied by auxiliary small spark discharges between pins situated adjacent to the main discharge that are connected in series with the peaking capacitors (Fig. 8.3).

Sometimes electrons supplied by separate corona discharges situated near the cathode are used for preionization. X-rays penetrate several metres through the gas and can be used to preionize uniformly much larger volume discharges than is possible with UV photons that are absorbed after travelling only a few millimetres. X-rays produced by a separate electron beam source pass into the gas through a thick metal foil or the cathode electrode. Only lasers that produce high pulse energies employ X-ray preionization and, because of the increased complexity, they tend to be more expensive and less common than UV preionized devices.

In modern excimer lasers the active lifetime of the electrical components such as capacitors, resistors, thyratrons, saturable inductors, electrodes, etc. should approach a billion (10^9) pulses which, with the laser running continuously at 100 pulses/s every day for 8 hours, will take ≈1 year to accumulate.

Table 8.3 Maximum performance of commercially available excimer lasers

	F_2	ArF	KrCl	KrF	XeCl	XeF
Wavelength (nm)	157	193	222	248	308	351/353
Pulse energy (J)	0.06	0.8	0.2	1.5	2	0.7
Peak power (MW)	3	30	10	50	50	30
Average power (W)	3	60	10	150	200	70
Pulse duration (ns)	10–30 (up to 200 possible on XeCl)					
Beam divergence (mrad)	2–10					

Note: The figures given are not achievable simultaneously with a single device.

8.1.5 Performance

Peak performances of various excimer lasers available commercially are shown in Table 8.3. The laser pulses usually last 10–30 ns and contain energies and peak powers of up to 2 J and 50 MW respectively in the wavelength range 157–353 nm.

Average powers of up to 200 W can be purchased either with a combination of relatively low pulse repetition rate and high pulse energy, or with a high (up to 1 kHz) repetition rate and low pulse energy. Figure 8.4 shows a photograph of a 150 W, 300 pps industrial excimer laser. While a few hundred watts is one to two orders of magnitude smaller than the powers produced by the largest industrial infrared Nd:YAG and CO_2 lasers, the short burst of the radiation produced by a pulse of excimer laser light leads to many interactions with materials that are unique and quite different from the way longer wavelength lasers interact.

By way of contrast, at the forefront of excimer laser development in scientific laboratories, average powers of over ≈1 kW in high-repetition-rate devices, and single pulse energies of ≈10 kJ having a peak power of $\approx 10^{12}$ W focusable to an intensity of 200 TW/cm^2 have been achieved. The latter performance was obtained from a large KrF laser machine used for inertial confinement thermonuclear fusion studies.

8.1.6 Optics

The excimer laser cavity consists usually of a plane-parallel resonator whose windows also serve as mirrors formed from polished UV-transmitting materials such as fused silica, crystalline CaF_2 or MgF_2. One of these has a highly reflective (≈100%) aluminium or dielectric coating on its rear surface. The substrate protects the coating from the harsh corrosive halogen gas environment. The output mirror is an aligned uncoated window with a two-surface Fresnel reflectivity of ≈8%. Prolonged exposure to intense

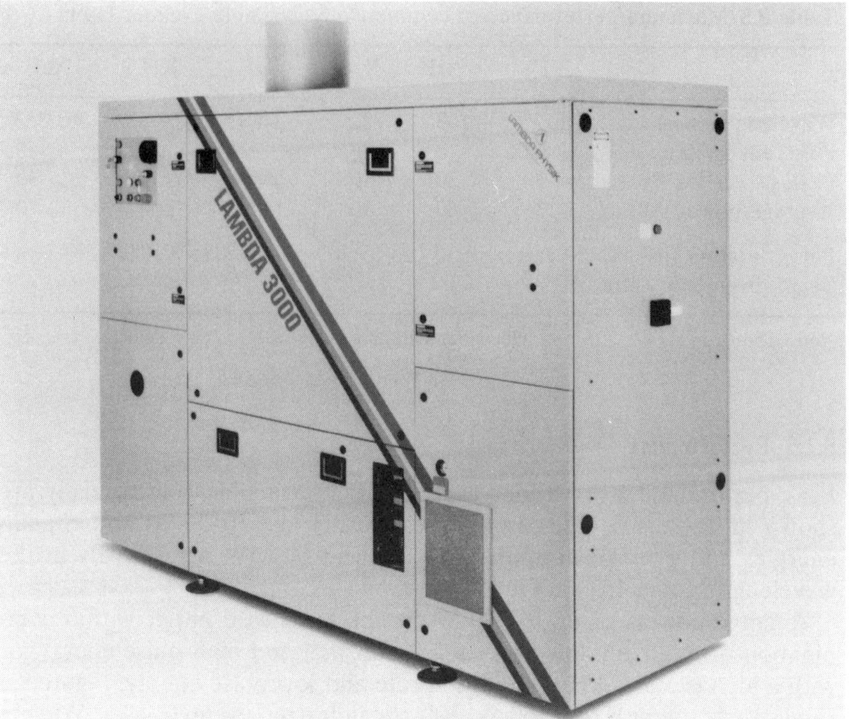

Fig. 8.4 150 W, 300 pps industrial XeCl laser. Photograph courtesy of Lambda Physik GmbH, Gottingen, Germany.

short-wavelength ultraviolet light often leads to the build up of lattice defects in the band gap of the mirror substrate materials decreasing their transmission in the UV. A good indication of such colour centre formation is when the window substrate fluoresces white, blue, green or red as the laser beam passes through it. MgF_2 is the material least susceptible to the build up of colour centres and most manufacturers now use this material for window and mirror substrates. If dielectric coatings are not used then the optics are common for all of the excimer laser transitions and a wavelength change is effected by only a simple change of the rare or halogen gas constituent of the fuel gas mixture.

Firing of the laser creates debris ablated from the electrodes and preionizer pins in the form of dust that coats the internal laser window and mirror. Gases such as hydrofluoric acid (HF) and ozone (O_3) along with the presence of intense UV light can also etch the mirrors so that periodically they must be removed and cleaned, polished or replaced. Such a cleaning procedure is usually necessary after $\approx 10^7$–10^8 pulses and in modern excimer lasers by the use of isolating gate valves can be carried out quickly without replacing the gas mixture.

The spectral gain bandwidth of $\geq$100 cm^{-1} of excimer lasers is so broad that, when using a plane-parallel stable resonator, up to a million axial and transverse modes oscillate simultaneously along their length of 80–150 cm and across their aperture of 1–3 cm^2. The spatial and temporal coherent properties of excimer lasers are more similar to the street lamp line sources of mercury or sodium rather than to the more classical types of laser such as He-Ne or Ar^+. Thus light interference effects such as fringe formation, holographic recording and laser speckle (interference caused by scattering from roughened surfaces), that are the hallmark of most laser sources, are often unobservable with beams produced by excimer lasers.

As shown by the beam profiles in Fig. 8.5, excimer laser beam shapes are determined by the discharge aperture size and are typically rectangular with aspect ratios of 2–3:1 and dimensions of (2–3) × (0.5–1) cm with highly divergent angular beam spreading of 2–10 mrad. To produce more collimated **spatially coherent** and hence more focusable beams from excimer lasers, confocal positive branch unstable resonator cavities are used in the reverse Cassegrain telescope type of arrangement shown in Fig. 8.6(a). When using magnifications of between × 10 and × 15, beam divergences can be reduced an order of magnitude to ≈200 µrad in beams that have ≈80% of the energy of what would be produced by the stable resonator. Instead of allowing them to build up from spontaneous emission, initially seeding such resonators as in Fig. 8.6(b) with radiation derived from a second excimer laser that can have an energy as small as ≈10 µJ, further improves the divergence close to the limit set by diffraction of light from the laser aperture *a* (diffraction limited divergence angle $\theta \approx \lambda/a$). When focusing the radiation from such a low divergence injection-seeded laser, exceedingly high intensities approaching 10^{14} W/cm^2 can be produced at the focus of a 10 cm focal length lens.

By placing frequency-selective elements such as Fabry-Perot etalons, diffraction gratings or prisms in the optical cavity, excimer lasers can be made **temporally coherent** by forcing them to operate with much narrower linewidths. A linewidth reduction of ≈600 to ≈0.1 cm^{-1} is readily achievable with methods that can then be used to tune the output wavelength across the broad-gain bandwidth of the laser. While such line narrowing is achieved usually at the sacrifice of laser power, by amplifying the output with, or injection seeding a second laser, as illustrated in Fig. 8.6(b), highly temporally **and** spatially coherent laser beams of the type required for performing holography can be produced in pulses having close to 1 J of energy.

8.1.7 Gases

The laser vessel is prefilled with a mixture typically consisting of a 4–5 mbar pressure of halogen gas F_2 or HCl, a few tens to hundred mbar of the working rare gas (Ar, Kr or Xe), and then pressurized to a total pressure

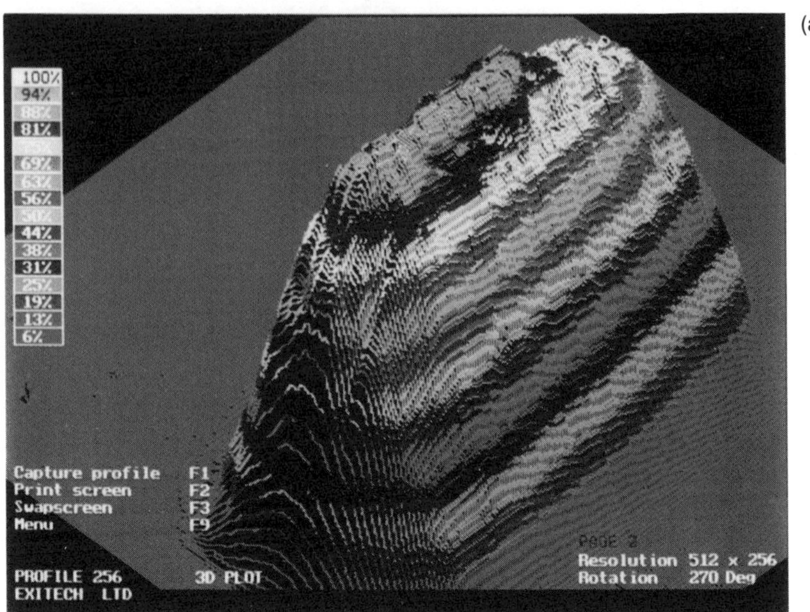

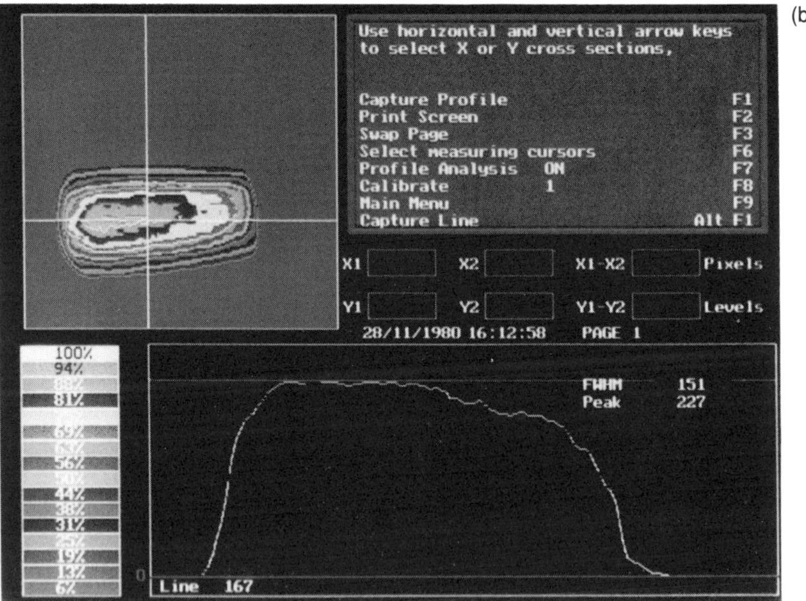

Fig. 8.5 Spatial profile of a typical excimer laser beam showing (a) 3-D isometric plot contours, (b) beam footprint and cross-section. Photograph courtesy of Exitech Ltd, Oxford, UK.

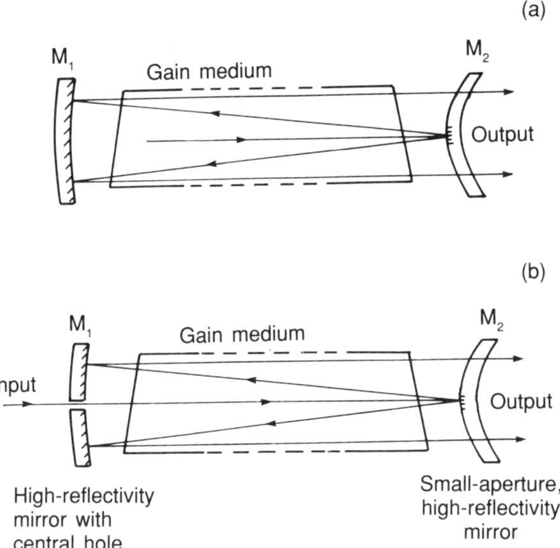

Fig. 8.6 Unstable resonators: (a) Cassegrain positive branch, (b) injection-seeded cavity.

of 2–4 bar with either He or Ne buffer gas. This working gas mix is then sealed and during operation is circulated by a blower in a closed loop through the electrode gap and a water-cooled heat exchanger. Dust which can scatter the laser beam and reduce its energy is removed by passing the gas through either integral electrostatic or particulate filters. Cryogenic trapping is also often used to remove UV-absorbing or other impurity gases which build up as a result of halogen reactions, repeated firing of the discharge and the presence of intense UV radiation.

The gases fluorine and hydrochloric acid in the fuel–gas mixture are extremely reactive and corrosive to the surfaces of most materials. Usually as a result of a seal failure, pressurized vessels containing them are prone to leaking after prolonged use (most excimer laser users are familiar with the distinct pungent 'swimming-pool' type of smell of fluorine gas!). Fortunately, the nose can detect this gas at extremely low concentrations of ≈1 p.p.m. To reduce the dangers associated with their high toxicity and corrosivity as well as the possibility of self ignition upon contact with fresh metallic surfaces, halogen gases suitable for laser use should always be purchased in a cylinder diluted with 95% helium. Such a mixture is then suited for filling the laser using the manufacturer's recommended gas cylinder regulator.

Prolonged exposure of the surfaces of many materials to halogen gas leads to the formation of a relatively unreactive skin. Such surface passivation coupled with a judicious choice of benign material (metals like nickel, aluminium, copper or brass, insulators such as fluorinated or chlorinated plastics – polyvinylidene fluoride (PVDF – Kynar) and polytetrafluoroethylene (PTFE – Teflon) – or ceramics like alumina), allow the sealed gas in the chamber to remain unreacted and suitable for laser use for a period of days or even weeks in the case of XeCl.

Due to the reaction and loss of the halogen fuel as well as the creation of new molecules and dust that absorb and scatter the laser light, firing of the discharge during laser operation causes the gas to degrade more rapidly than by passive reactions. In Fig. 8.2(a) we show how the continuous spectrum of a KrF laser has been eaten away by absorption bands of molecules such as HF, O_2 and CF_2 that are created by repetitively pulsing the discharge – hydrogen and oxygen are always present in excimer lasers due to the presence of small gas leaks and impurity traces of water vapour. While filtration techniques are used to minimize the build up of dust and cryogenic trapping can be used to remove some of the chemical contaminants, eventually the laser mix will cease to produce a useful laser power and will have to be replaced. While this usually means stopping operation to evacuate and refill the vessel, replenishment can be computer controlled and when operating the laser continuously can be performed gradually without stopping.

The lifetime of an excimer gas mixture is like asking 'how long is a piece of string?'. It depends on many parameters such as the transition used (ArF, KrF, XeCl, etc.), the number of laser pulses fired, the laser repetition rate, the cleanliness of the gas supply and the laser optics, the use of a cryogenic trap and the make of the laser. The performance often quoted is in terms of numbers of laser pulses and is in the range 10^6–10^7 for ArF and KrF and 10^7–10^8 for XeCl. A several-fold increase in these numbers can be achieved by continuously circulating the gas in a closed loop through a cryogenic trap to remove impurity gases and to reduce the buildup of debris on the laser window and mirror. This type of lifetime performance is achieved only when operating continuously – normally at the maximum repetition rate of the laser (50–500 pps say). When operating in this mode, gas running costs on KrF and ArF for example can be as low as ≈$1–2/hour when using a cryogenic processor and ≈$10–30/hour when not. When not operating in a continuous manner fewer pulses will be accumulated before gas replenishment is required – in the limit that no pulses are fired then the static lifetime of the gas in the tube of several days to weeks means that the shot lifetime of the laser is zero and gas running costs are substantial! Neon-buffered mixtures are between 2 (XeCl and XeF) and 15 (ArF) times more expensive to fill the laser with than are helium ones. Operational experience of high repetition rate excimer lasers used on industrial production lines has shown running costs to be between $25–30/hour and dominated by spare part and equipment costs.

8.2 EXCIMER LASER BEAM DELIVERY SYSTEMS

Of equal importance as the laser to most applications is the integration of a satisfactory beam delivery system that can tailor the beam shape and size, deliver it to the processing area and mechanically or automatically maneouvre it around safely and reliably. Because of the applications areas involved and the different beam shapes, sizes, and wavelengths of excimer lasers, to use the beam efficiently often challenges the optical and mechanical designer to produce quite different beam delivery systems from those encountered with other types of laser sources.

8.2.1 Beam homogenizers

A beam profile produced by a typical commercial excimer laser is shown in Fig. 8.5. The computer 3-D isometric plot of spatial variation of power has been recorded by a beam profiler with a CCD or Si diode array camera placed directly in the beam. It can be seen that the beam is rectangular in crossection with an aspect ratio of ≈2–3:1. In the narrower dimension the intensity profile is approximately Gaussian while in the wider one it is flatter and more like a super Gaussian shape. The variation of intensity (uniformity) in such a beam is at best between ±20–25% over an area that contains about 80% of the energy of the pulse. For most of the application areas discussed in Chapter 9, such an inhomogeneous beam is unsatisfactory for processing purposes. Depending on the application uniform top hat, Gaussian, hollow, spatially multiplexed or other types of tailored profile are usually preferred.

While a beam can be made more homogeneous by aperturing its size, energy is thrown away in the process and for highly inhomogeneous beams is a very inefficient method. Several optical devices for efficiently homogenizing the output from excimer lasers have been tried. Most rely on splitting the beam up spatially into smaller components and then recombining and overlapping them further downstream from the device. Thus, while the divergence of the beam is increased, at some plane and over a depth of field determined by the divergence of the beamlets the beam can be highly uniform.

As is discussed in Chapter 9 section 9.4.1, an homogenizer that consists of a silica diffuser plate and a rectangular kaleidoscope rod working at relatively low throughput efficiency is used in photolithographic step-and-repeat machines to produce the highly uniform beam required to illuminate the mask in the fabrication of integrated circuits. Multiple fly's eye lenses that overlap and focus different transverse parts of the beam on top of each other can produce homogeneous beams at the focus but the lenslet arrays are difficult to fabricate in fused silica and are expensive to buy. For efficient operation they must be antireflection coated for

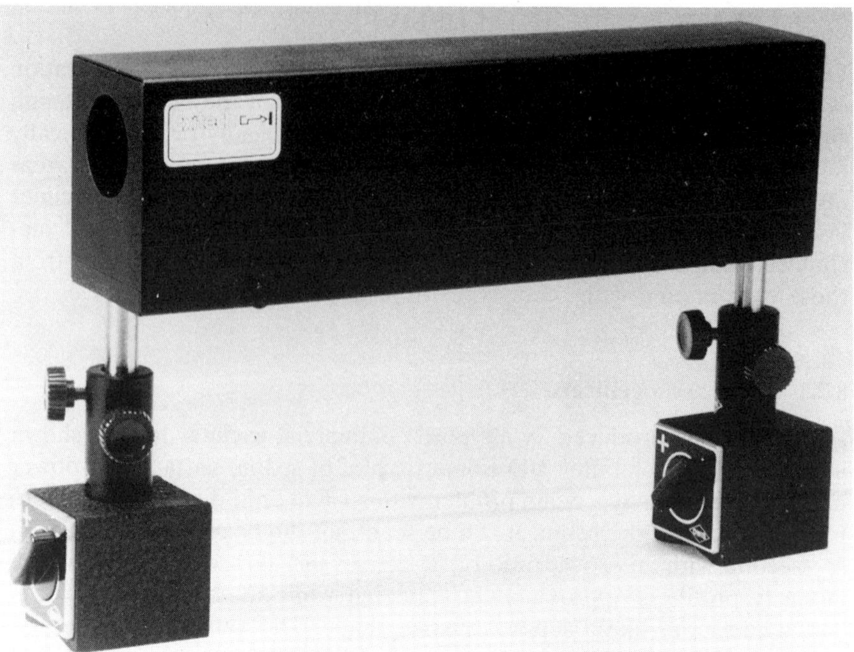

Fig. 8.7 Excimer laser beam homogenizer. Photograph courtesy of Exitech Ltd, Oxford, UK.

ultraviolet light. Large-diameter single or bundled fused silica optical fibres can also be used to scramble the beam but tend to be lossy in the UV and cannot operate with energetic beams without incurring damage. They also produce a widely diverging beam at the output that is difficult to transport efficiently on to the workplane. Reflective devices that operate by multiple grazing reflections between two nearly parallel mirrors also tend to be lossy and damage at high powers.

Optical elements such as random phase plates that introduce a spatially random π variation in phase across the beam can be used to improve the uniformity of the beam in the far field. Computer generated holographic elements and spherically aberrating telescopes can produce uniform beams from specific (e.g. Gaussian) input beam profiles. The device shown in Fig. 8.7 uses a bi- or quadraprism to spatially multiplex and reconstitute the beam and is an efficient method for homogenizing the near field beams from excimer lasers. Cylindrical lenses reshape the beam and, as shown in the sketch in Fig. 8.8, refractive bi- or quadraprisms symmetrically split, invert and overlap it on itself. Depending on the laser used, square top-hat beam profiles that are more suited for processing applications can be produced with efficiencies of >90% and uniformities better than $\pm 4\%$ over >90% of the 10–20 mrad beam spread (Fig. 8.9). Rather than being highly

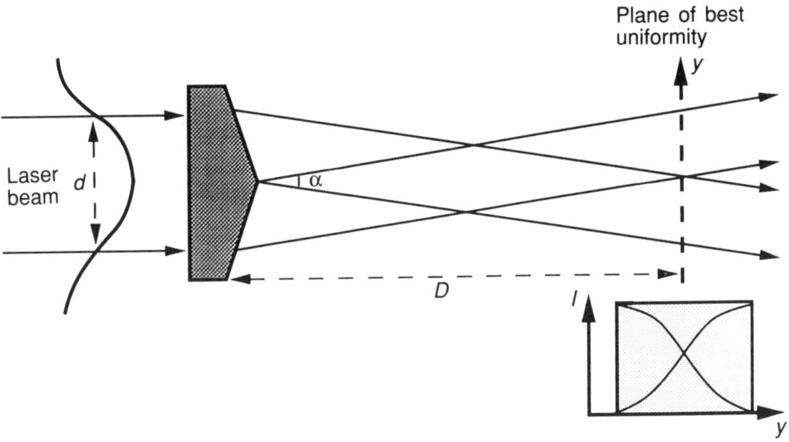

Fig. 8.8 Principle of operation of the homogenizer shown in Fig. 8.7. The plane of best uniformity is the position at which the beamlets overlap and are inverted.

uniform, the device relies only on the original laser beam being symmetric in orthogonal transverse directions. Masks or apertures can be placed in the plane of best uniformity which, with simple spherical lenses, can then be image relayed on to the processing plane.

8.2.2 Beam multiplexers

To split up (multiplex) a laser beam into many smaller ones each with similar intensities, may be more appropriate for applications that require several processing steps to take place simultaneously. Such multiplexing can be performed either by passing the beam through a succession of beamsplitters to divide up its amplitude, or, by using small mirrors, prisms or lenses, to spatially split up its wavefront. Amplitude division of a high-power laser beam requires the use of many expensive coated beamsplitters that are prone to laser damage and different split ratios are needed to produce beams that have similar intensities. The relatively large dimensions of excimer laser beams are readily amenable to using wavefront division for beam multiplexing. Variations on the multiple-prism type of device described above for homogenizing the beam can also be used to divide it up spatially into smaller beamlets each having similar intensities. In Fig. 8.10 such a device is used to split the beam up and simultaneously drill 16-hole arrays in a moving webb of thin plastic foil.

8.2.3 Image projectors

Processing of material with excimer lasers is usually most efficiently carried out using an appropriate mask inserted into the beam. Reduced images

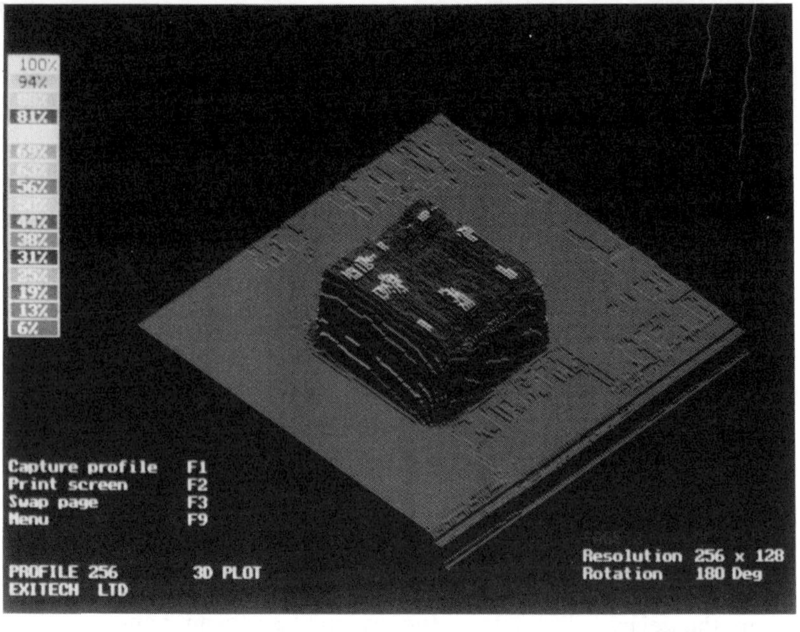

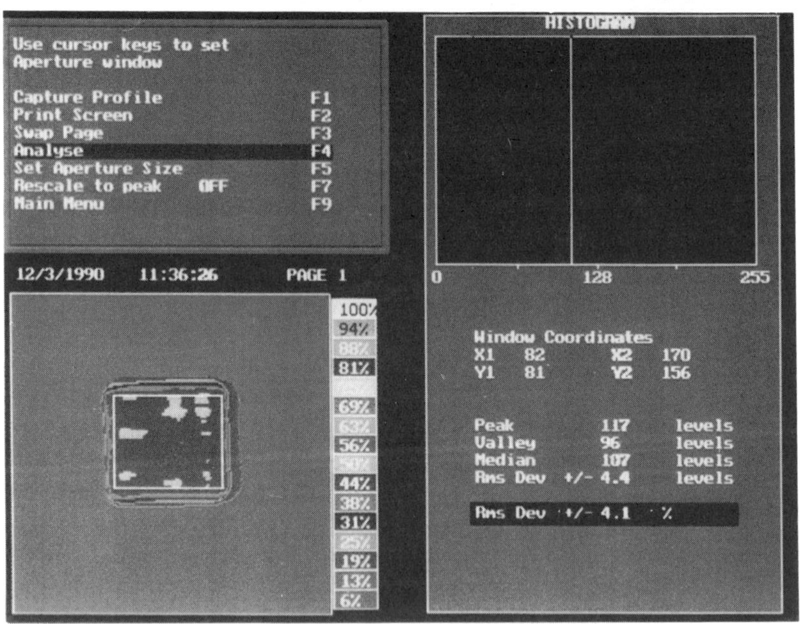

Homogenized beam top hat 10x10 mm
energy fluence rms dev ±4.1%

Fig. 8.9 Excimer laser beam profile obtained using the homogenizer in Fig. 8.7. For this case the histogram shows that within the rectangle which contains ≈90% of the energy of the beam there is an intensity uniformity of ±4.1%.

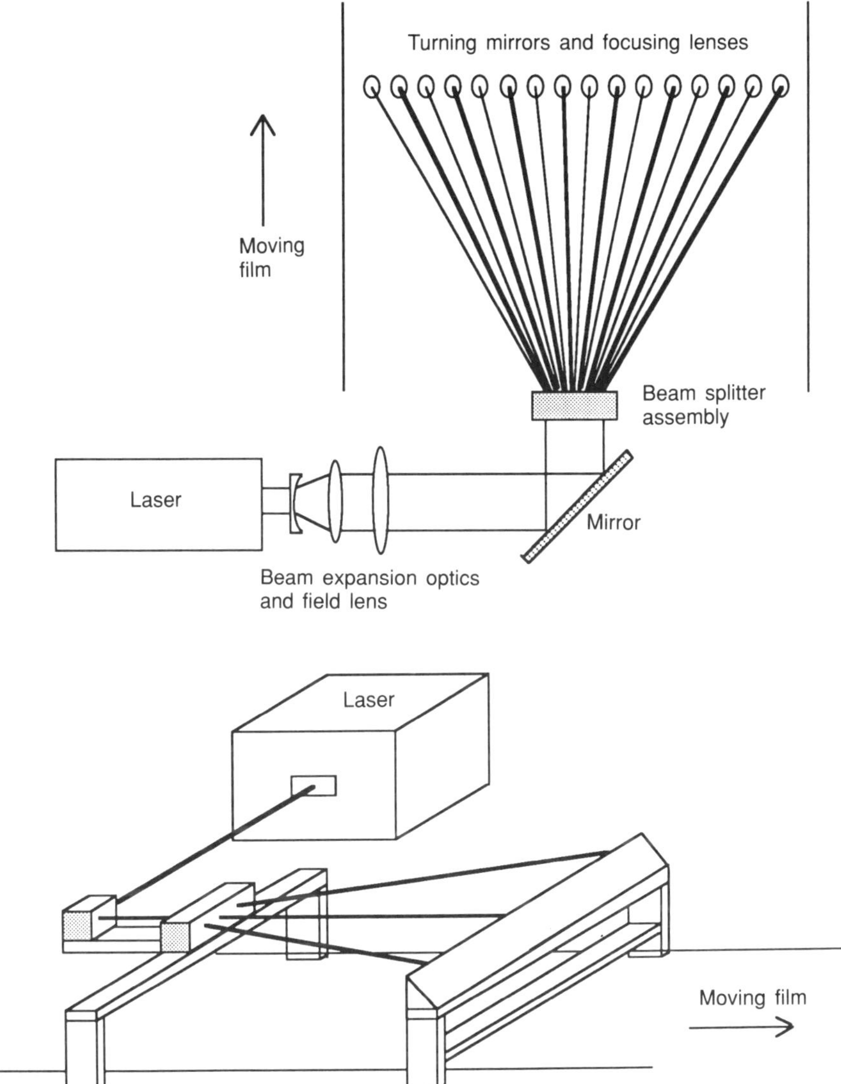

Fig. 8.10 For simultaneous processing with many beams – such as might be useful for multiple hole drilling in moving films for example, the beam can be spatially multiplexed with small prisms or mirrors to produce beamlets having similar intensities.

of the mask patterns are then relayed with lenses on to the workpiece. The resolution and size of the image required determines the complexity of the imaging lens. For images of a few millimetres containing features of ≥5 μm

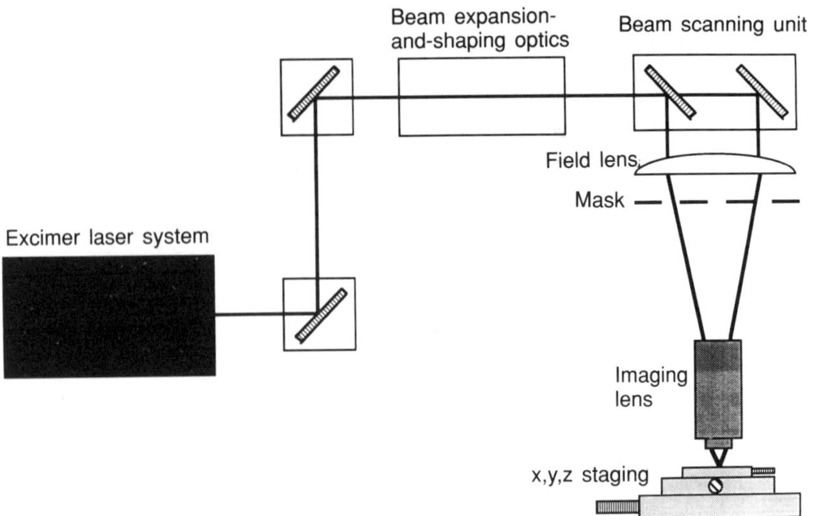

Fig. 8.11 Excimer laser image projector incorporating beam homogenizer, imaging and field lenses and unit for scanning the beam across large masks.

in size, simple singlet or doublet spherical lenses of fused silica are sufficient for imaging. On the other hand, to achieve resolutions of ≤1 μm over this size requires the use of multicomponent lenses sometimes fabricated from different transmissive materials to correct for image defects such as astigmatism, coma, and spherical and chromatic aberration.

One method of treating material over areas of >0.5 cm^2 at energy densities in excess of 1 J/cm^2 with a laser that produces only a fraction of a joule energy is to use cylindrical lenses to reshape the beam into a stripe that is then scanned across the mask as the laser is repetitively pulsed. As shown by the schematic in Fig. 8.11, such scanning is achieved conveniently using a motor-driven mirror and illuminating the mask through a large field lens so that the transmitted diffracted light remains within the aperture of the fixed imaging lens. To cover large areas on a workpiece, another option is to scan the mask and workpiece in synchronism across the fixed laser beam.

The photograph in Fig. 8.12 shows an excimer laser image projector consisting of beam homogenizer for uniform illumination of the mask, beam profiler for mask plane monitoring, mask changer, and field and imaging lenses integrated together in one device. Zooming of the imaging lens allows different sizes and intensities of the uniformly illuminated image to be produced at the workpiece. This projector can be used for microhole drilling, marking, annealing, demetallization and photoresist exposure over variable field sizes and exposure densities.

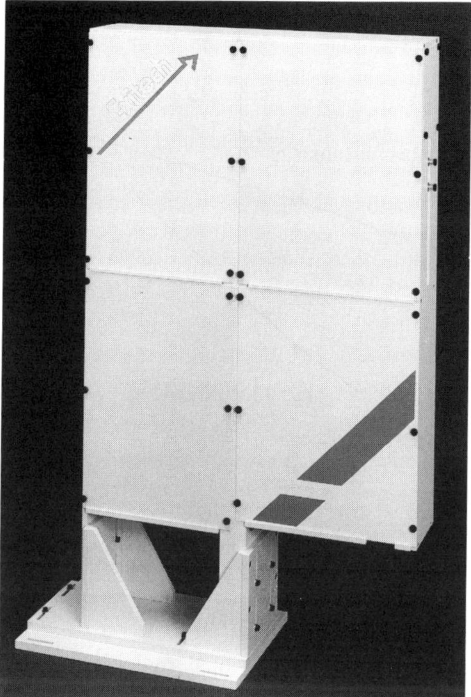

Fig. 8.12 Excimer laser image projector incorporating beam homogenizer, profiler for monitoring the beam at the mask plane and imaging and field lenses that produce variable demagnifications of the mask at the image plane. Photograph courtesy of Exitech Ltd, Oxford, UK.

8.2.4 Fibres

The ability to deliver energetic laser pulses to a location somewhat remote from the laser with a high degree of user controlled flexibility is clearly the great advantage offered by fibre optic beam delivery systems. Unfortunately, the attenuation suffered by the beam as it propagates through silica fibres and the damage effects incurred at high transmitted powers become more severe at shorter wavelengths. At low laser powers in the ultraviolet, linear fibre transmission losses have been found to increase as $1/\lambda^5$, presumably because of intrinsic absorptions and the $1/\lambda^4$ Rayleigh scattering from small inhomogeneities in the silica fibre. As shown in Fig. 8.13, at 248 nm, a ≈3.5 m length of silica fibre will transmit only 50% of the radiation launched into it. For high-power laser radiation shorter than ≈300 nm propagating through fibre lengths of a few metres, a major optical loss process is two photon nonlinear absorption in the silica. For example, the

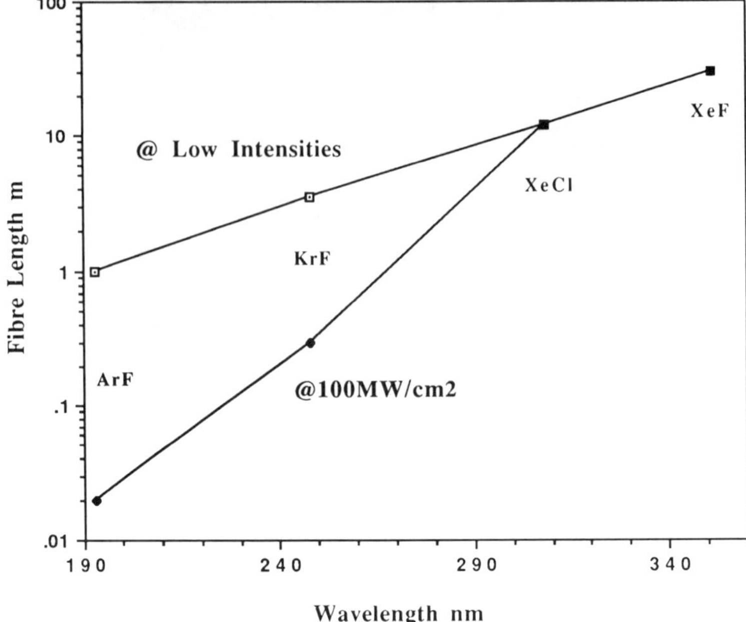

Fig. 8.13 Lengths of silica fibre that produce 50% transmission of the input radiation for different excimer laser wavelengths.

transmission of the 3.5 m length of silica fibre is reduced to ≈10% when the intensity at 248 nm is increased to ≈100 MW/cm^2.

Irreversible damage to the input face of the fibre occurs at a single pulse threshold fluence that decreases as $\approx\lambda^4$. For a cleanly cleaved fibre and a 248 nm pulse of ≈30 ns duration having a top hat intensity distribution without hot spots, damage to the input face occurs at fluences as high as ≈35 J/cm^2. To operate within a safety margin of about half the damage threshold then with a Gaussian beam at 248 nm an energy of ≈6 mJ can be safely launched into a 400 μm core diameter silica fibre. Using lengths of 0.6 and 3.5 m, fluences of ≈2.8 and 0.5 J/cm^2 respectively would be present at the exit of this fibre. The latter fluence is barely sufficient for efficiently cutting most plastic or biological materials, so only the minimum length of fibre required for the application should be used.

Novel fibre launching geometries that employ tapers, index matching liquids and lensed ends are being tested as methods of increasing the damage threshold for launching high-power beams. The damage threshold scales as the square root of the pulse duration and so is higher for longer pulses. For 300 ns pulses at 308 nm, ≈19 mJ of energy can safely be launched into the fibre. Since the threshold fluences and cutting rates for most plastic

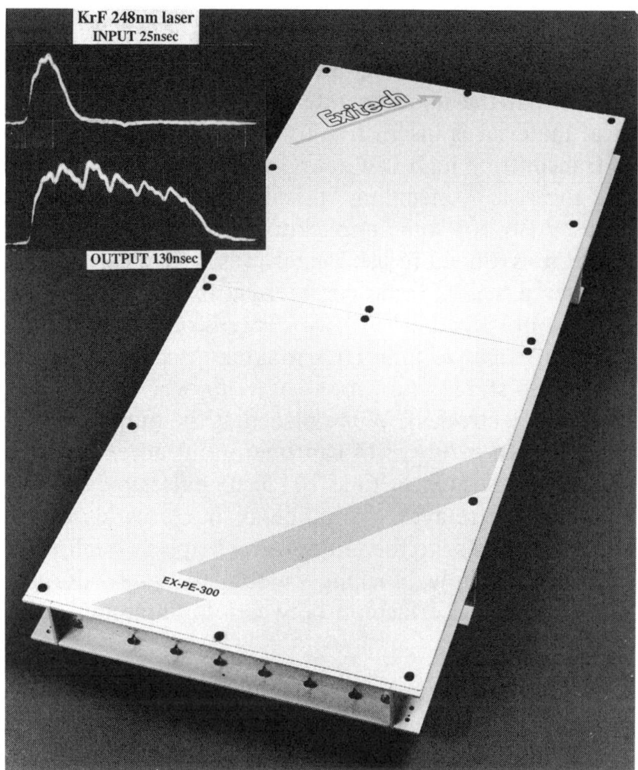

Fig. 8.14 Excimer laser pulse extender. In the example shown, a 25 ns duration KrF laser pulse is transformed into a 130 ns pulse with ≈65% energy conversion efficiency. Photograph courtesy of Exitech Ltd, Oxford, UK.

and biological materials are relatively insensitive to the laser pulse duration, long pulses can be used to transmit high UV energies down optical fibres to increase cutting speeds at the exit and reduce the risk of damage to the input face. Excimer lasers that produce such long pulses directly or pulse extenders that stretch the pulses from short pulsed devices can be purchased from several manufacturers.

The buildup of colour centres and microscopic damage sites in the fibre due to the propagation of intense UV laser beams currently limits the useful life of the best silica fibre to a few tens of thousand pulses at 308 nm and about an order of magnitude less at 248 nm. Because of two photon and long-term induced absorption losses, the excimer laser of choice for fibre optic delivery is XeCl. Future developments of fibre optics for use in the UV may allow for more practical delivery of high powers at the more useful polymer cutting wavelength of 248 nm.

8.2.5 Pulse extenders

As mentioned above, in the range of tens to hundreds of nanoseconds it has been found that the ablation rate and threshold for etching plastics and biological material is insensitive to the excimer laser pulse duration. Thus, when transmitting high UV energies down a silica fibre for processing purposes, the risk of incurring optical damage to the fibre is reduced for long pulses. Also, for some applications, for example to laser–surface interactions, there is a need to use longer pulses than can be generated by the excimer laser directly. Some commercial excimer lasers can produce longer pulses but only at the XeCl wavelength of 308 nm (fluorine gas is much more electronegative than HCl, making discharges in mixtures containing it much less stable and prone to arcing when long voltage pulses are applied to the electrodes). When placed at the output of the laser, the pulse extender shown in Fig. 8.14 can operate at all excimer laser wavelengths to efficiently stretch in time 10–20 ns pulses by over an order of magnitude. This image-relaying optical delay line provides an output beam size and divergence similar to the input, with energy throughput efficiencies exceeding 50%. By simply blocking mirrors the pulse duration can be easily 'tuned' in a stepwise fashion between its minimum and maximum values.

8.2.6 Articulated arms

As with other types of laser, excimer beams can be delivered flexibly to a working area using appropriately coated turning mirrors, prisms and lenses. Because of its relatively large divergence, to transport an excimer laser beam over distances of greater than about a metre without it becoming too large and unwieldy, requires the use of converging lenses or mirrors to image relay planes close to the laser. In all other respects the mechanics of the articulated arms are similar to those used with other lasers.

8.2.7 Processing systems

An example of a complete processing system that incorporates and integrates the laser, homogenizer, beam delivery, imaging optics, beam profiling, sample viewing and workpiece handling is shown in Fig. 8.15. Such a machine is currently used in production and preproduction application areas in microelectronics for via hole drilling, mask and chip repair, wafer annealing and marking, and for fabricating devices such as biomedical sensors and ink jet printer nozzles. The laser beam train and processing areas are fully contained in a radiation-tight vented interlocked enclosure that exhausts and filters the fumes created during processing.

Fig. 8.15 Excimer Laser micromachining workstation. Photograph courtesy of Exitech Ltd, Oxford, UK.

FURTHER READING

Bachmann, F.G. (1990) Industrial laser applications, *Applied Surface Science*, **46**, 254–263.

Excimer Lasers, 2nd edn (1984), *Topics in Applied Physics*, vol. 30 (ed. C.K. Rhodes), Springer-Verlag.

Various authors (1986) Excimer lasers and optics. *SPIE*, vol. 710.

Various authors (1988) Excimer lasers and applications. *SPIE*, vol. 1023.

9

Excimer lasers: current and future applications in industry and medicine

M.C. Gower

9.1 THE INTERACTION OF EXCIMER LASER RADIATION WITH MATERIALS

While excimer lasers do not produce as high average powers as longer wavelength infrared lasers such as Nd:YAG and CO_2, in principle their shorter wavelength allows their beams to be focused to smaller spots and obtain high intensities at the workpiece. Nearly all of the applications of excimer lasers realized in industry today and those potential ones on the horizon, make use of the unique manner in which these lasers interact with materials. In many cases the processing of material and the types of structures that the excimer laser is capable of producing are not even possible to achieve by other manufacturing techniques. It is these applications on which we will concentrate. Shown in Table 9.1 are the binding strengths of some of the most common chemical bonds that hold atoms together to form molecules.

These energies are compared with the energy carried by a single excimer laser ultraviolet photon and are in the range 3.5–7.9 eV. Each photon carries sufficient energy to break apart many of the molecular-based materials commonly found in nature as well as those chemically synthesized by man particularly the 3.5 eV C═H bond crucial to all organic materials. This is extremely significant since to break material apart with the light from longer wavelength lasers such as Nd:YAG ($\lambda = 1.06$ μm, $h\nu \approx 1.2$ eV) and CO_2 ($\lambda = 10.6$ μm, $h\nu \approx 0.12$ eV) requires the absorption of many photons to heat the material hot enough to cause its molecules to dissociate. As we shall see later, such high temperatures usually lead to many undesirable effects such as combustion, charring, melting, flow and boiling of surrounding unirradiated material. Illuminating many substances in solid, liquid or

Laser Processing in Manufacturing. Edited by R.C. Crafer and P.J. Oakley.
Published in 1993 by Chapman & Hall, London. ISBN 0 412 41520 8

Table 9.1 Strengths of some common molecular chemical bonds compared with excimer laser photon energies.

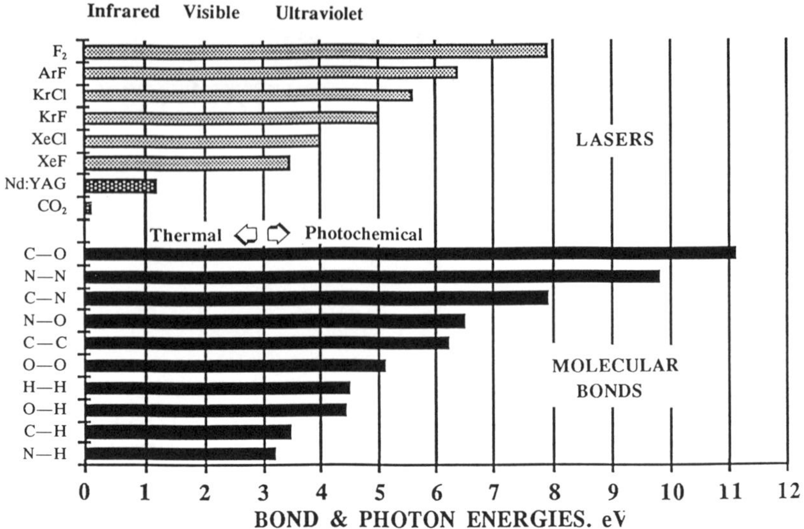

gaseous form with excimer laser photons often substantially modifies them by stimulating a change to either their chemical nature or structural form.

The ability of the excimer laser to induce photochemical change to a gas is best illustrated by sniffing, in the vicinity of a 193 nm ArF laser beam as it passes through air, the pungent ozone (O_3) smell characteristic of seaside locations. Each photon in the laser beam has sufficient energy to break apart individual oxygen molecules (O_2) in its path, some atoms of which then recombine with other molecules to form ozone. The smell is not evident near the beams of the longer wavelength excimer lasers since their photon energies are less than the 5.1 eV molecular oxygen bond and so cannot individually break it.

The structural modification that can be caused to solid materials by exposure to UV excimer laser light is easily demonstrated by holding various plastics in front of the unfocused beam and observing the flame and hearing the crack created as the top layer of material is removed. In this case the photons interact with the material by individually breaking the polymer chemical bonds that hold it together and give it its mechanical integrity. Plastics consist of long chain high molecular weight polymer molecules, each comprising many tens of thousands of atoms (usually H,C,O and N) repeated in monomer units of a few tens of atoms. The short burst of excimer laser light is absorbed in a shallow ($\leq$10 μm) depth of these materials and rapidly breaks the relatively weak polymer bonds (binding energies <0.5 eV) at the surface. As depicted in Fig. 9.1 in the confined

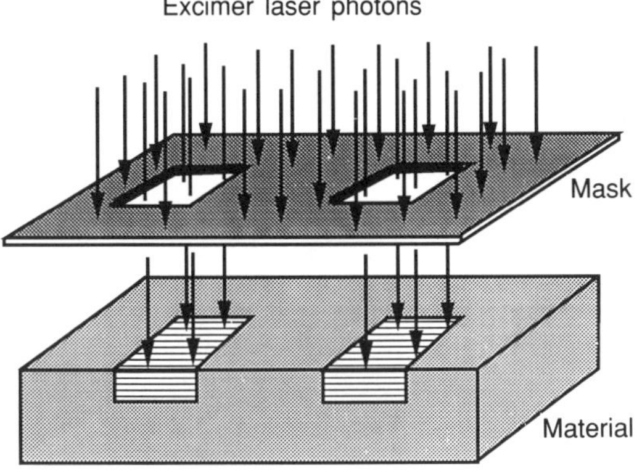

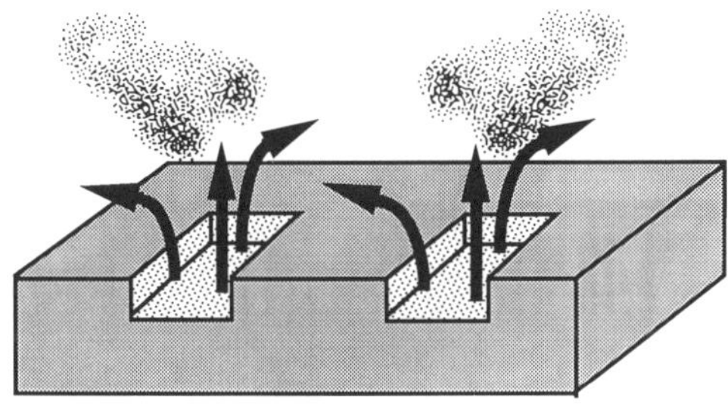

Fig. 9.1 The short-duration UV excimer laser pulse rapidly breaks chemical bonds in a polymer within a restricted volume to cause a mini-explosion that ejects material.

illuminated volume this bond breaking leads to a rapid rise in the local particle number density so, like in an internal-combustion engine, there is a corresponding rapid rise in pressure. Instead of pushing a piston, the pressure is released as a shock wave that ejects material fragments of gases and monomers at high speeds from the irradiated zone. Often the fragments themselves are excited and fluoresce in a flame-like manner to give the appearance of a plume of light emanating from the exposed site.

Because there is very little excess heat transferred to the surrounding material, this process of 'ablative photodecomposition' or 'photoablation' can be used to great affect in finely patterning objects made from materials such as plastics, paper, ceramics, glasses, crystals, composites and biological tissue. The precision of excimer laser cutting is spectacularly demonstrated in Figs 9.2 and 9.3 which show scanning electron microscope (SEM) photographs of a human hair into which ≈40 μm square holes and notches have been machined with an ArF laser. There is no evidence of thermal damage to the surrounding unirradiated tissue. Further studies have shown that individual biological cells can be cut in half with an excimer laser

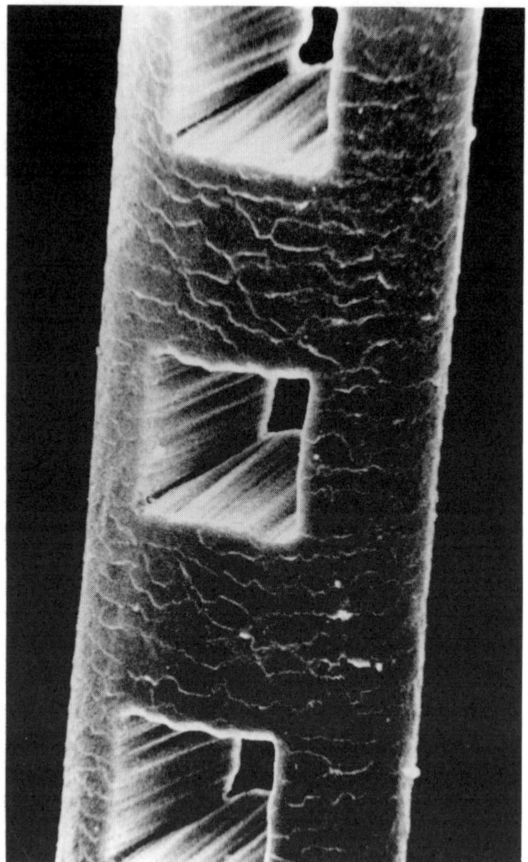

Fig. 9.2 40 μm square holes drilled in human hair with an ArF excimer laser.

without apparently incurring damage to the remainder of the cell. Another example of this cutting precision is shown in Fig. 9.4, where a 200 μm diameter plastic hollow microballoon has had a 40 μm hole drilled through it by the laser.

Unlike many applications of lasers, photoablative etching makes simultaneous use of several of the unique properties of laser light. The process relies on the production of an intense short burst of light to create a rapid rise in pressure. High intensities are required to interact with many molecules simultaneously. The UV wavelength is necessary for ensuring that the chemical bonds are broken directly. Most molecular-based materials are highly absorbing in the UV so that dissociation occurs in the top ≈1 μm layer of material. The narrow wavelength spectrum of the laser offers the possibility of selectively breaking specific chemical bonds leaving others in

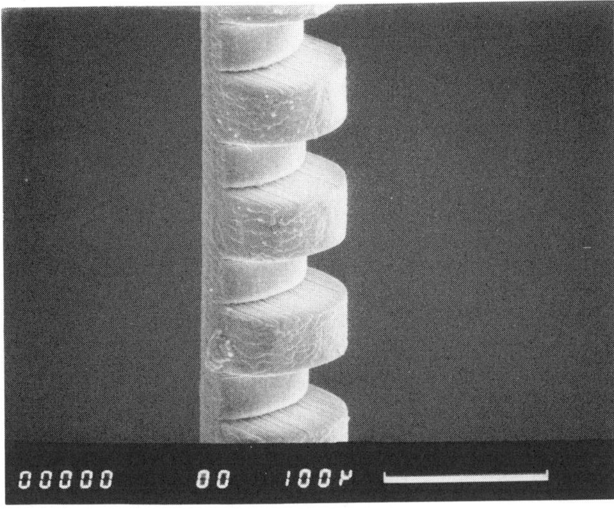

Fig. 9.3 40 μm notches machined in a human hair with an ArF excimer laser.

the materials untouched. The spatial coherence of the laser source allows the beam to propagate over large distances, to be focused to small spots or to be used to relay images of intricately patterned objects placed in its path without incurring appreciable loss in intensity. Specific chemical reactions and material modifications stimulated by excimer laser light which are or might be utilized in industry are the subject of the following sections.

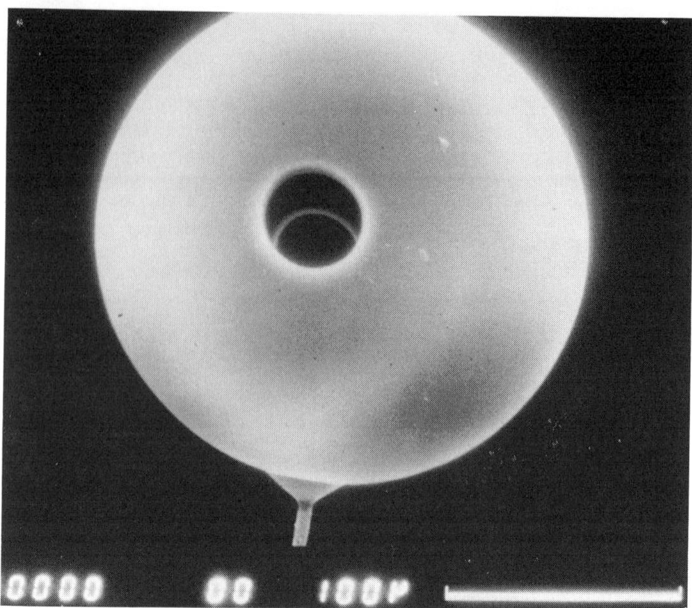

Fig. 9.4 40 μm round holes drilled through a 200 μm diameter hollow plastic microballoon with an ArF excimer laser.

9.2 LASER MICROMACHINING OF PLASTICS, METALS, CERAMICS, GLASSES AND COMPOSITES

9.2.1 Plastics

In Fig. 9.5 we show an example of an array of blind square holes which have been drilled in a sheet of polyethylene terephthalate (PET or Mylar) by excimer laser photoablation using ArF, KrF and XeCl lasers. When using the shorter wavelengths of 193 and 248 nm the wall edges of the cuts are extremely well defined. At the bottom of the cuts the ≈0.2 μm features arising due to diffraction of light from the edge of the contact mask used to define the holes is easily resolved. However, at the longer XeCl laser wavelength of 308 nm removal of material occurs by heating of the irradiated site. As can be seen in Fig. 9.5, this leads to melting and flow of the base polymer at the bottom and edge of the cut. Figure 9.6 shows further examples of patterns cut in polymers. In contrast to the contact mask used to obtain the cut in Fig. 9.5 these etches in polycarbonate (Lexan) were produced by projecting with a fused silica lens the image of a mask pattern placed in beam. Again the quality of the cut and the well-defined wall boundary are excellent.

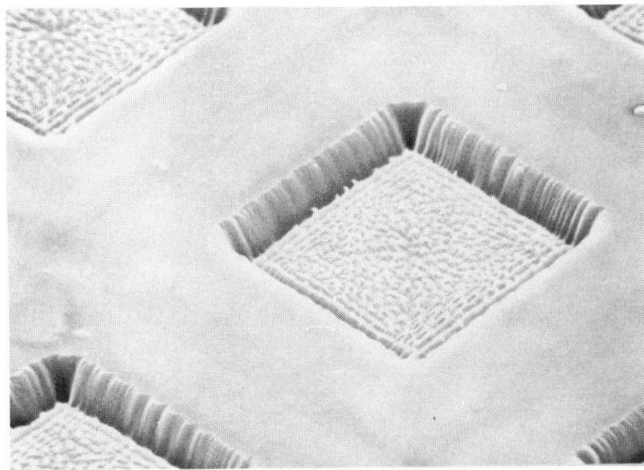

Wavelength=193 nm
Fluence=0.62 J/cm^2

Wavelength=248 nm
Fluence=0.72 J/cm^2

Wavelength=308 nm
Fluence=1.4 J/cm^2

75 μm square patterns

Fig. 9.5 75 μm squares etched in PET to similar depths with ArF, KrF and XeCl lasers.

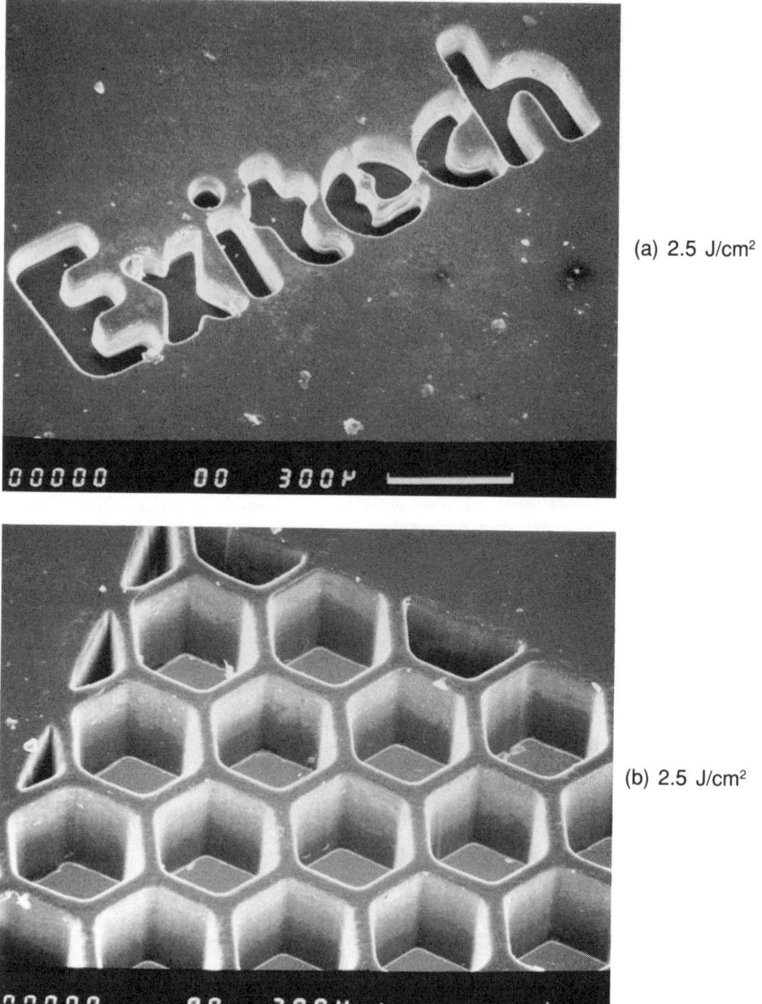

Fig. 9.6 Etches in polycarbonate at 248 nm using a KrF laser and image projection of mask patterns.

The measured cutting rates as a function of laser fluence for the plastics polyimide (Kapton) and PET (Mylar) are shown in Figs 9.7 and 9.8 for the three most common excimer laser wavelengths. Depending on the laser wavelength and energy fluence, cutting rates of several tenths of μm per pulse are typical for most polymers once a well-defined threshold is exceeded. Below this threshold fluence no etching occurs, because insufficient polymer bonds are broken for the induced pressure rise to overcome

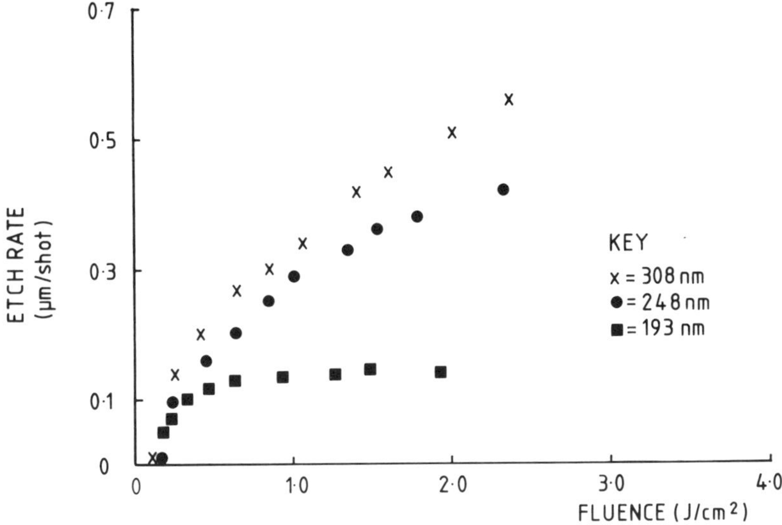

Fig. 9.7 Rate for etching polyimide per pulse versus fluence with the three principal excimer lasers.

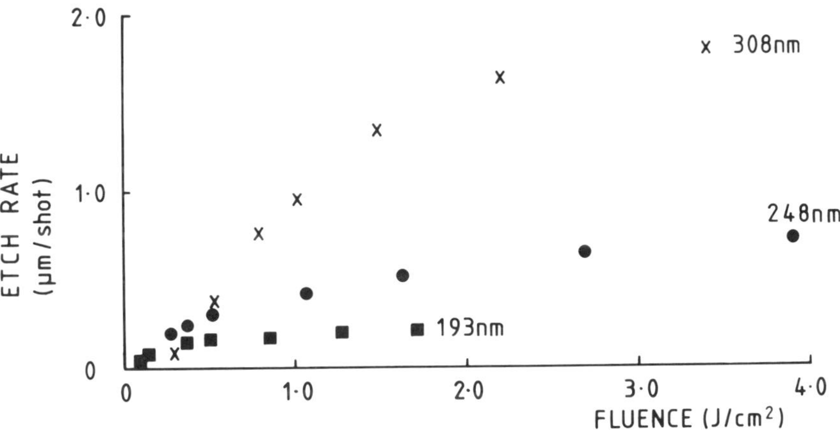

Fig. 9.8 Rate for etching PET per pulse versus fluence with the three principal excimer lasers.

the mechanical integrity of the material. Saturation of the cutting rate at higher fluences is also a common feature of the photoablation process and occurs when there are sufficient photons to break all of the polymer bonds within the illuminated volume defined by the area and photon penetration depth into the material. If one photon breaks one polymer bond in a material having a density of ≈1 gm/cm^3 and a monomer molecular weight of 100, then the volumetric removal rate at saturation is ≈200 nl/J. In the Appendix at the end of this chapter the threshold fluence and the maximum linear and volumetric etch rates for excimer laser cutting of variety of materials are given.

While the small depth removed for each pulse may seem to be a major drawback for using the excimer laser for machining, it allows for an unprecedented accuracy in the precision of cutting and drilling operations. By simply counting the number of laser pulses the depth of a cut can be controlled precisely to within a fraction of a μm. Furthermore, the laser energy fluence required to exceed the threshold for etching in polymers is very modest (≈100–500 mJ/cm^2) and is similar to the unfocused single-pulse fluence produced directly at the laser output. Thus, unlike when cutting material with focused beams from infra-red lasers, relatively large areas of 1–2 cm^2 at the workpiece can often be processed with a single laser pulse from the excimer. By treating the material with many patterns at once rather than serially processing it, as is done in more conventional laser cutting, scribing or drilling operations, this type of processing is most efficiently utilized in a parallel type of operation. The relatively low processing speed caused by the small removal rate of material in a particular region is then often more than made up for by the high degree of parallelism which can be implemented with large area coverage.

If we assume that over 1 cm^2 a depth of ≈0.3 μm of material (≈30 μg) can be removed by each laser pulse at a fluence of ≈600 mJ/cm^2 and that we have a laser which produces 100 pulses/s with an average power of 70 W, then a maximum of ≈3 mg/s of plastic can be removed. Hence to remove 1 kg of material from the surface would take about a week, working 12 hours every day. Since similar fluences are used when single-pulse marking or otherwise surface treating polymers with excimer lasers and providing the sample can be moved at speeds of ≥1 m/s so that each pulse treats a fresh surface, then in the same time an area of ≈3000 m^2 can be covered.

As mentioned above and as shown in Fig. 9.9, the most effective method of processing material with excimer lasers is to replicate into the workpiece the required pattern that is contained in a mask placed in the laser beam. Masks can either be placed in close contact with the sample to be treated (Fig. 9.9a), or a lens can be used to project on to the workpiece an image of the mask (Fig. 9.9b).

With contact replication the mask receives the same laser fluence as the workpiece. To avoid damage caused by the high laser intensity, free-standing or conformal metal (copper, nickel, etc.) masks should be used. Conformal masks can be deposited conveniently on to plastics by standard lithographic techniques. Sacrificial masks of a similar material to the workpiece that are deliberately ablated away by the laser can also be used etch patterns

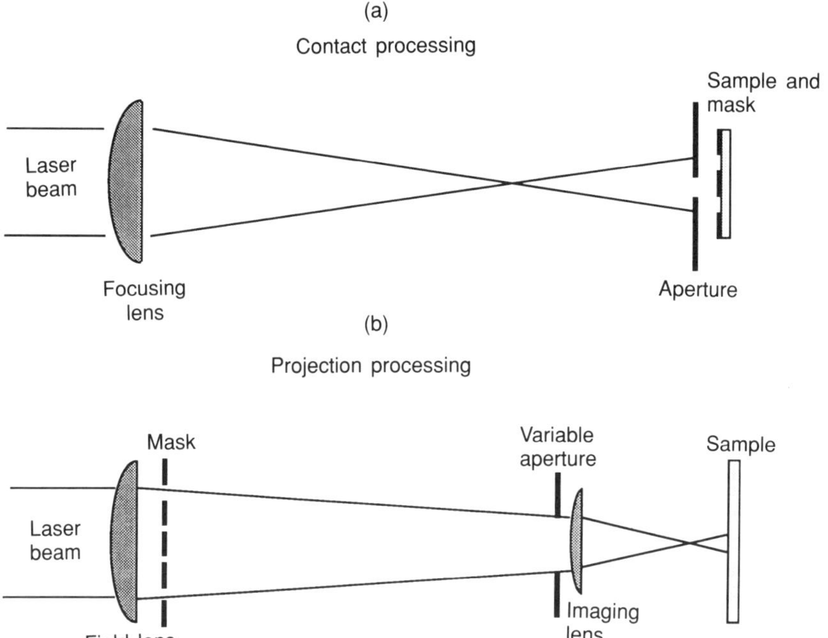

Fig. 9.9 Patterning of material is optimally performed with the excimer laser by using a mask containing the pattern and either (a) placing the two in contact, or (b) using a lens to image the mask on to the material.

differentially into or to shape the workpiece. Contact processing is relatively simple to set up and can produce exceedingly high resolution replication with minimum feature sizes approaching the wavelength of the source (the minimum feature size W is given by $W^2 \approx 0.8\lambda d$, where d is the mask/workpiece separation). Large areas can be covered by repetitively scanning the mask/workpiece combination across the beam between the limits of the full area to be processed. Alternatively the beam can be rastered across the combination using motor-driven steering mirrors. Debris often ends up clogging the mask in contact processing and it and the sample tend to damage as a result of the contact which must be exceedingly good to maintain the theoretical resolution. It is also exceedingly difficult to fabricate free-standing masks that contain structures having sizes of <5 μm.

While usually more difficult to set up initially, replication by mask imaging has several advantages over contact replication. Processing by projection is more easily automated for repetitive operations and the mask is usually well away from the workpiece so is less prone to contamination by processing debris. Since imaging optics can produce a demagnified image on to the workpiece, the job of making the mask is made much simpler

Fig. 9.10(a) A 0.3NA 10 mm field size 10:1 reduction lens consisting of six antireflection-coated fused silica components, used for high-resolution imaging of masks illuminated with a broadbandwidth KrF laser. Photograph courtesy of Rutherford Appleton Laboratory, UK.

x10 Image reduction, resolution ≈1.6 μm; 15 mm Field

Polycarbonate etched with KrF laser at 248 nm (no line-narrowing)

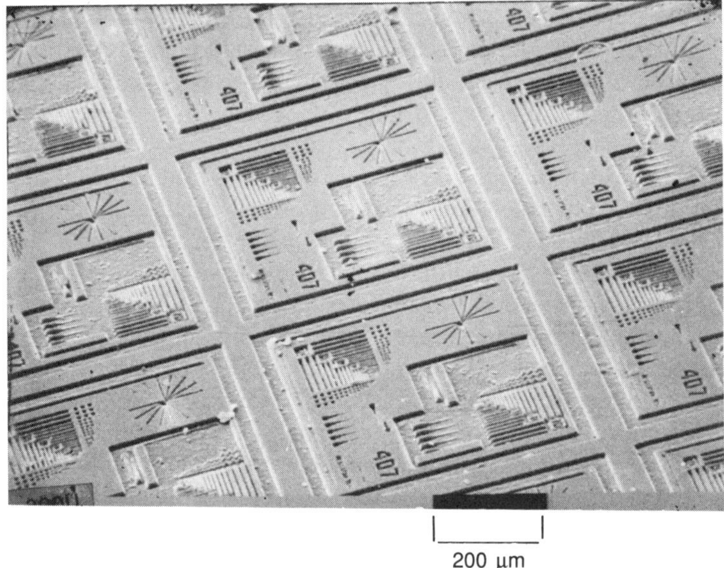

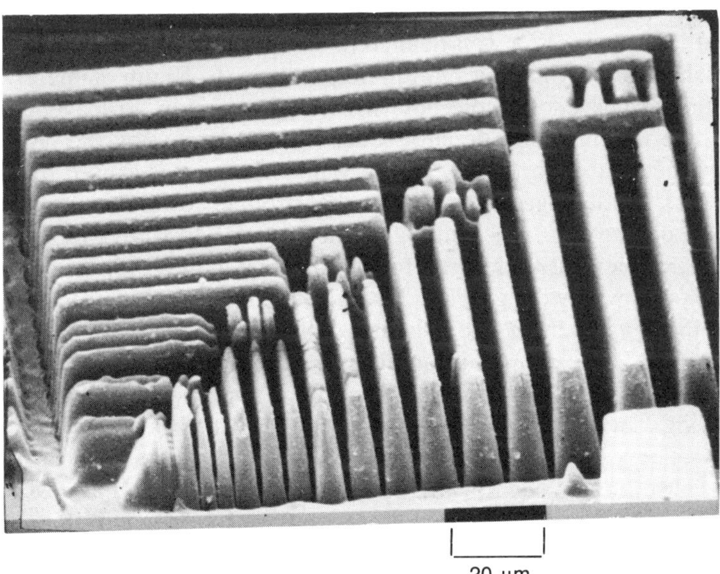

Fig. 9.10(b) A single image of a (multicomponent) resolution test mask etched into polycarbonate sheet using a lens similar to (a).

and the fluence on it is reduced by the square of the demagnification so laser induced damage is less likely to occur. Patterns over large areas on the workpiece can be produced by stitching together in a step-and-repeat fashion exposures of the same or different mask patterns. Masks can be fabricated accurately – chemically etched free-standing or electron-beam-written chrome lines on fused silica substrates are popular choices. The latter is the standard mask used by the photolithography industry for producing integrated circuits. The main drawback of imaging replication is that the precise plane which gives the best in focus image is often difficult to locate. Without auxiliary focusing aids, trial and error is used and once found it is accurately referenced and relocated in further processing.

For high-resolution diffraction-limited imaging, large numerical aperture lenses must be used and the distance over which the image is in focus is short – the resolution of an image $W \approx 0.8\lambda/\mathrm{NA}$ and its depth of focus $z \approx 0.8\lambda/\mathrm{NA}^2$ where NA is the numerical aperture of the imaging lens. Due to the many optical aberrations which must be corrected for in diffraction-limited imaging in the ultraviolet spectral region, the design and construction of suitable demagnifying lens combinations that have resolutions of less than ≈1 μm over field sizes larger than ≈10 mm is difficult and so they are expensive. In Fig. 9.10 is a photograph of a ×10 image demagnification multicomponent lens combination used for excimer laser micromachining of plastics in a step-and-repeat workstation.

Figure 9.11 shows that when using the excimer laser to drill holes at modest laser fluences, a well-defined hole taper is produced; in this case in a thin sheet of polyimide a 170 μm hole reduces to 50 μm diameter. If the material is thick enough the hole will eventually stop, to leave a conical blind hole which cannot be cut deeper without increasing the laser pulse fluence incident upon the workpiece. The hole taper is caused by diffractive effects at the edge of the pattern that produce lower fluences and etch rates in this region. Once initiated the taper is favoured by the lower fluence experienced at the wall for non-normal angles of incidence of the beam. As shown by the graph in Fig. 9.12, the taper angle becomes smaller as the incident laser fluence increases. This tapering can often be used to great effect. For example, in applications that require fluid flow through holes fabricated in thin membranes, tapering nozzles are favourable for producing laminar flow. When drilling holes in thicker materials total internal reflections of light at the walls within the tapered hole can eventually cause the fluence towards the edges to increase and a re-expansion and splaying of the taper is sometimes observed deeper into the material.

Depending on parameters such as the type of polymer, the laser wavelength and incident fluence, many other kinds of modifications to surfaces can be produced following exposure to excimer laser light. For example, the cone-like and roughened structures shown in Figs 9.13 and 9.14 can be produced at modest laser fluences in plastics such as PEI and

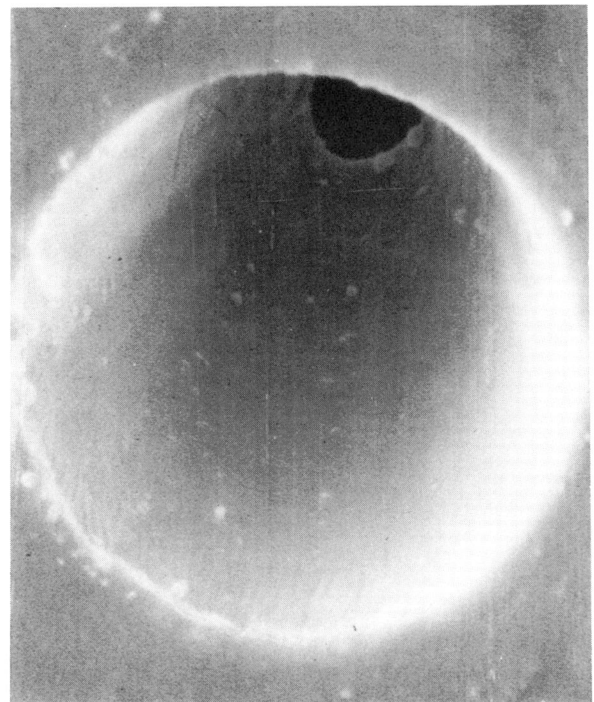

Fig. 9.11 A 40° tapering hole drilled in a 150 μm thick sheet of polyimide with a KrF laser at a fluence of ≈400 mJ/cm^2.

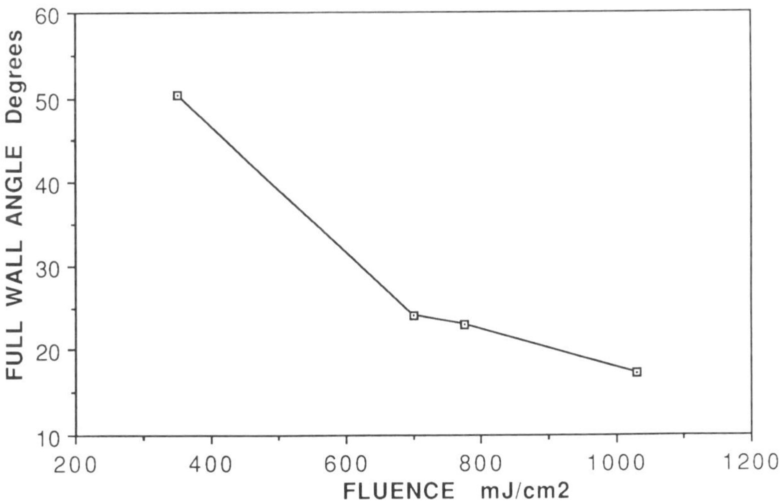

Fig. 9.12 Wall taper angle of holes drilled in PLA as a function of ArF laser fluence.

(a) Wavelength=248 nm
0.29 J/cm^2

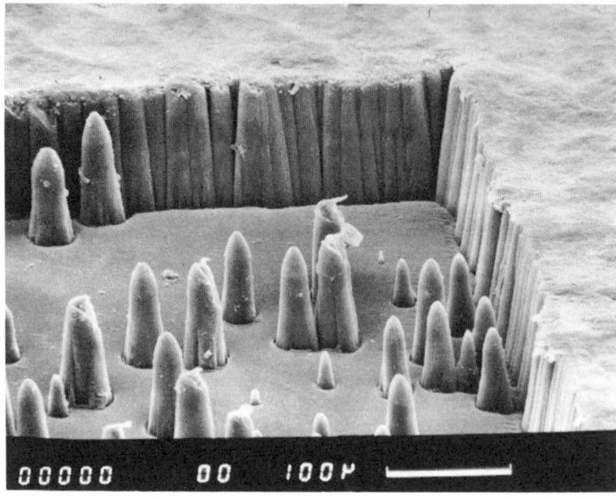

(b) Wavelength=193 nm
0.24 J/cm^2

Fig. 9.13 Cone-like structures etched in PEI with KrF and ArF lasers at fluences near the threshold for etching.

polyimide. Since similar structures can be induced by loading the base plastic with small particles of carbon, it is believed they are produced by small particulate impurities or redeposited debris on the surface of the material acting as miniature masks that prevent etching of material below. At higher incident fluences they are less evident because the particles themselves are blown away by the laser. On a microscopic scale such morphologies are extremely rough and have large surface areas which can enhance their adhesive bonding properties and be useful for applications in micropore filtration, surface catalysis and to reduce the specular reflection of light. Other examples of excimer laser induced surface modification are

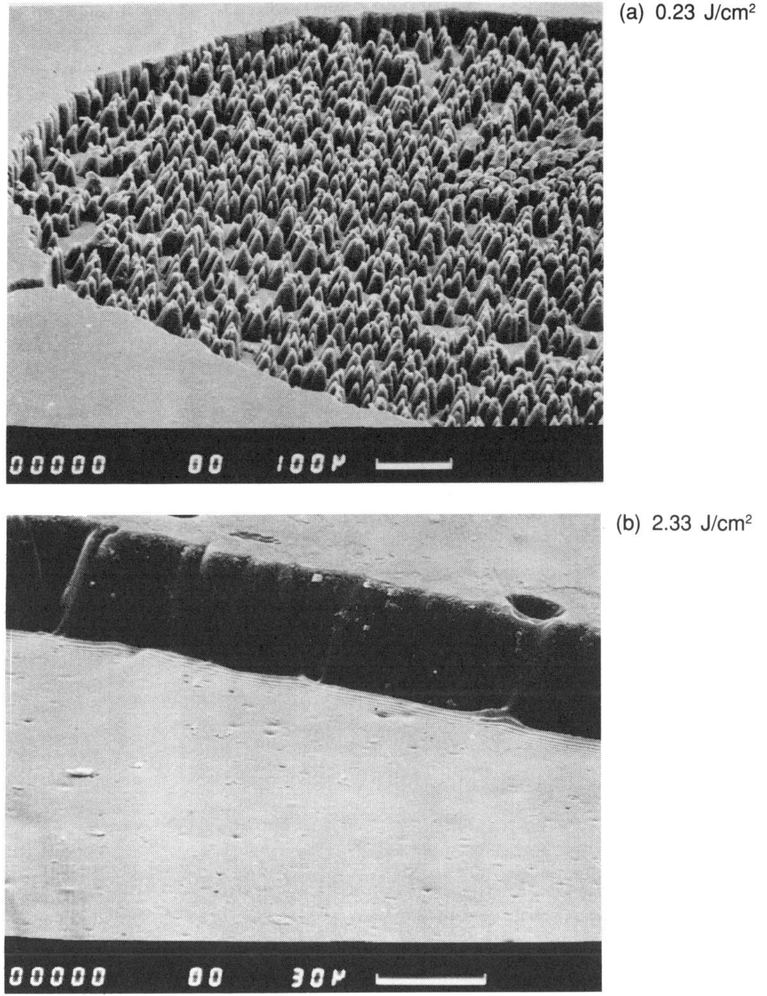

Fig. 9.14 Etches in polyimide with a KrF laser (a) just above the etching threshold, showing cone like structures, (b) smooth etching at a fluence ≈10 times the threshold for etching.

shown in Figs 9.15a and 9.15b for polycarbonate and polystyrene respectively. In these cases it appears that the processes used to fabricate the plastic (cast, rolled, extruded, polycrystalline, etc.) is partly responsible for creating the strange and beautiful structures produced.

Following excimer laser treatment of some plastics the electrostatic properties of its surface can be changed from being water-repelling hydrophobic to hydrophilic (wetted by water). The measured change in the

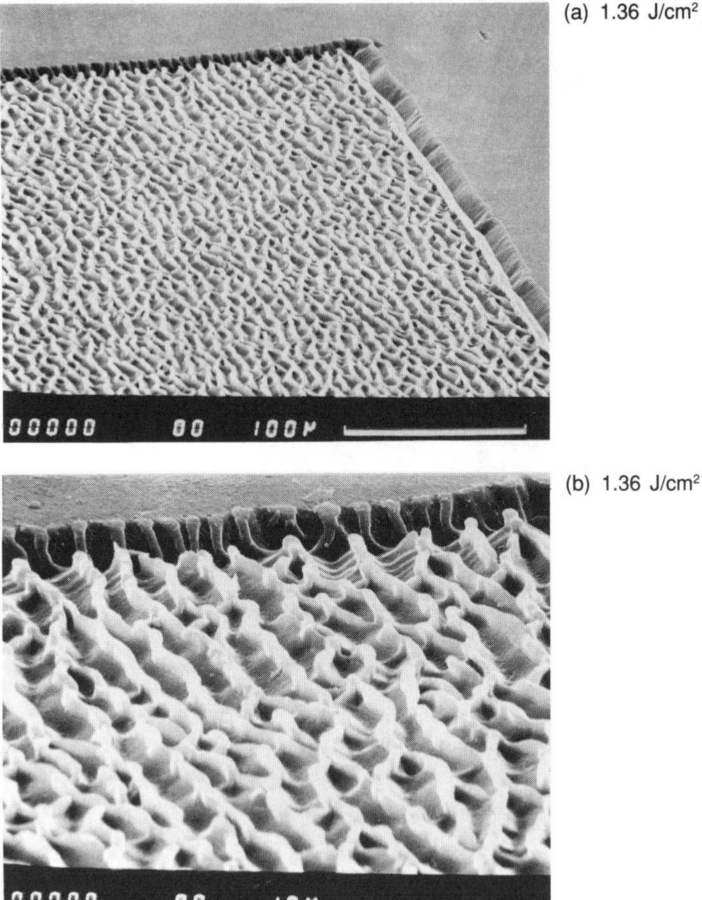

Fig. 9.15(a) Rough wave-like structures created by etching polycarbonate Makrolon foil with a KrF laser.

static surface contact angle for water droplets on three types of fibre composite surfaces is shown in Fig. 9.16 as a function of laser fluence. Notice that surfaces of the carbon fibre/epoxy and PEEK composites can be made hydrophilic by treating them with ~2 J/cm^2 of KrF laser radiation. Also, treating the PEEK/carbon fibre material at a lower fluence induces an increase in its hydrophobicity. The ability to manipulate the wettability of surfaces by excimer laser treatment could find wide applicability in the printing and adhesive industries. Because irradiation of surfaces by large patterned excimer laser beams can be carried out by contact or projection replication printing of a mask, it is easy to treat only certain predefined

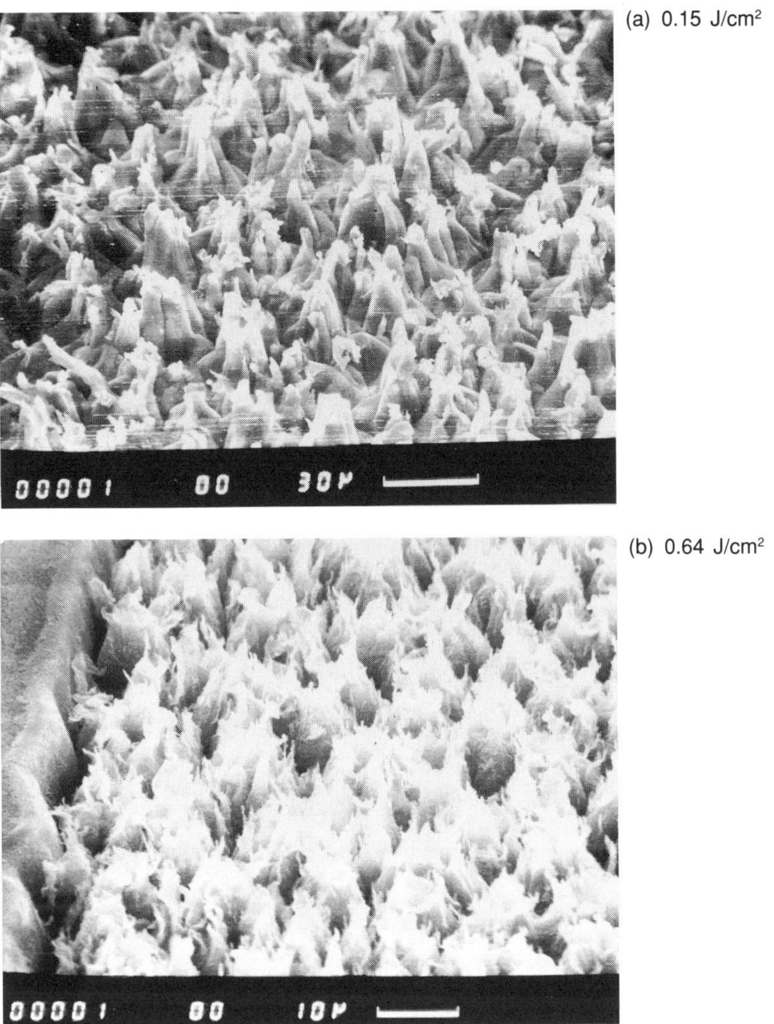

Fig. 9.15(b) Fluffy surface structures created by etching polystyrene with an ArF laser.

regions. For example, the adhesion of paint or dye to only the treated areas would be useful to a variety of industries.

Laser-induced surface modification of materials is likely to find widespread applications in the fabrication of devices such as micropore filters, membranes, surface catalysers, biological sensors and many other microstructures of interest to chemical, medical and biotechnologies. In Fig. 9.17 are examples of 1 mm and 100 μm hole arrays drilled with an excimer laser into a piece of card and a banknote respectively. The cellulose bonds in

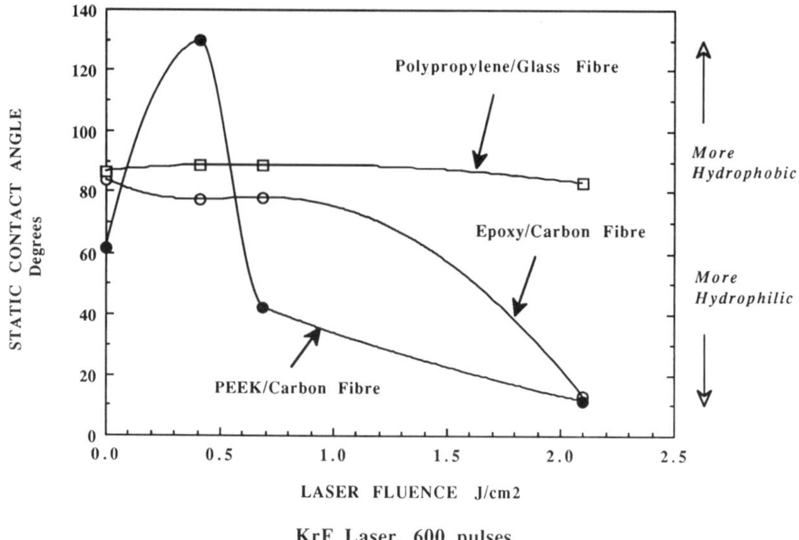

Fig. 9.16 Static contact angle of water droplet on KrF laser-treated surfaces. 600 pulses.

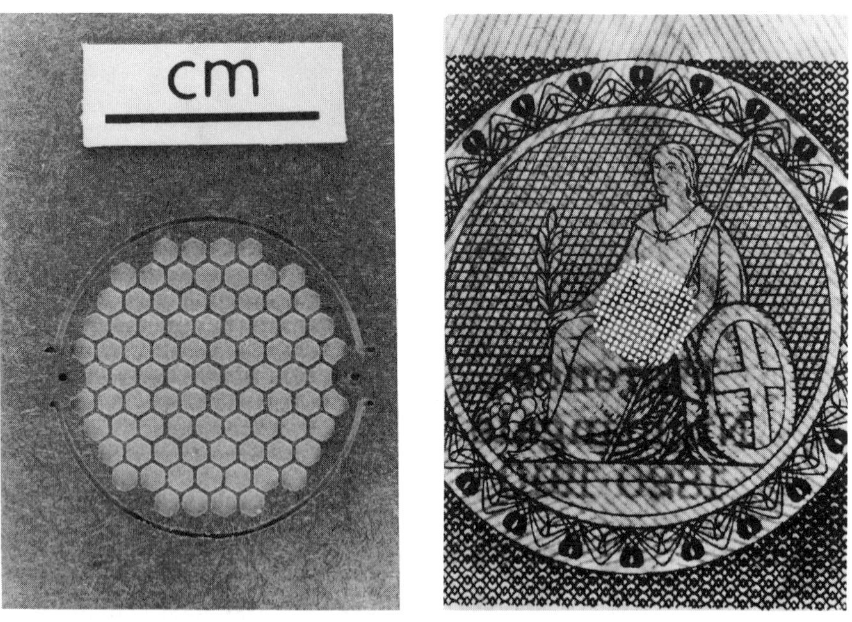

Fig. 9.17 Arrays of 1 mm holes drilled in a piece of card (left) and 100 μm holes drilled in a banknote with an ArF laser.

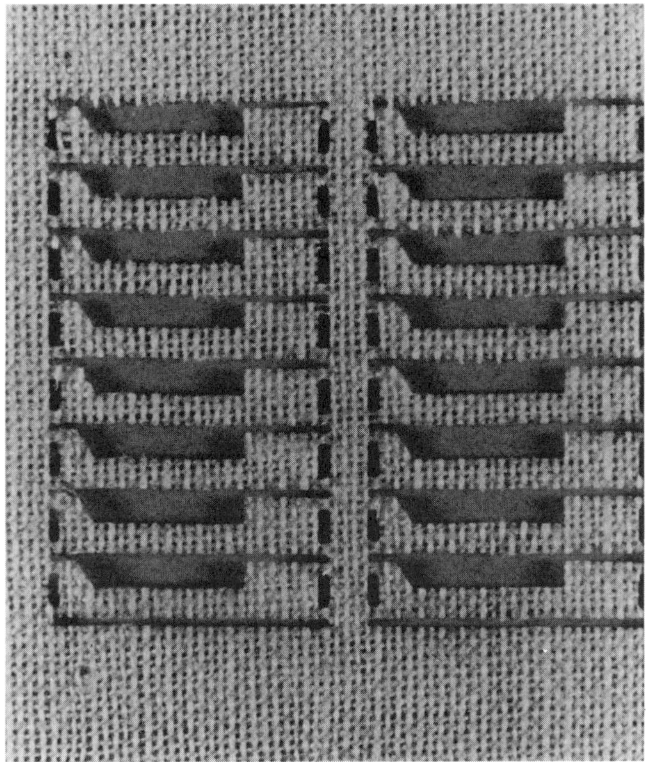

Fig. 9.18 Patterns cleanly machined by an ArF laser in woven cotton material.

paper or cardboard are readily broken by the laser to produce clean char-free holes. Since it would be extremely difficult to fabricate such high quality small holes by any other method this excimer micromachining could be applied to preventing counterfeiting of items such as banknotes and personal, legal and security documents.

Natural and synthetic woven fabrics such as silk, cotton, nylon and polyester can also be cut and machined very precisely with the excimer laser. As seen by the pattern cut in Fig. 9.18 there is no burning or fraying of the surrounding woven material. Exposure to excimer laser light can also change the static/antistatic properties of fabrics. Woven material can be cleaned of impregnations of dust-like particulate matter. The laser can also improve the adhesion of small sub-μm sized particles and pigment dyes to woven fabrics, which is of interest in improving filtration methods and fabric dyeing techniques.

Examples of hole and membrane patterns micromachined with a KrF laser by contact mask replication into thin films of polyimide and PET are

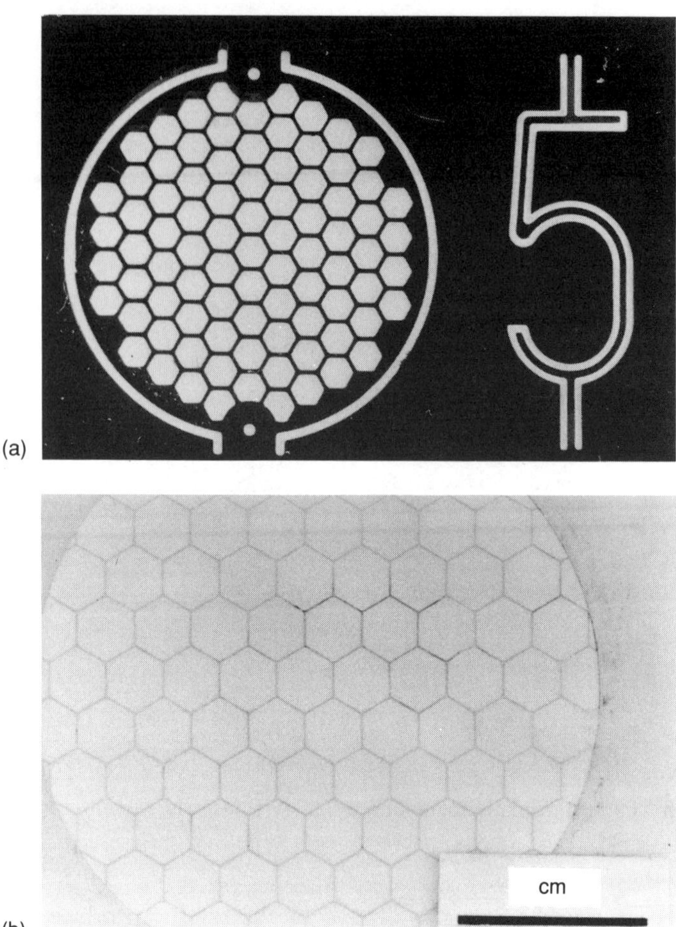

Fig. 9.19 Membrane hole arrays machined with a KrF laser in thin films of (a) polyimide, (b) PET.

shown in Fig. 9.19. The cutting is clean and precise while the piece itself retains much of its mechanical strength.

In Figs 9.20–9.24 examples of practical devices are shown that in some part of their fabrication use an excimer laser for micromachining. Figure 9.20 shows an example of a bilumen medical catheter that detects with electrical sensors the oxygen content of the blood of prematurely born babies. The hole at the side of the PVC tube through which blood is drawn is machined by KrF excimer laser. In this case the clean cutting capability of the excimer provides the necessary rigidity that prevents kinking and blockage of the tube when inserted into the child. Also shown in Fig. 9.20 is a sensor consisting of a 200 μm diameter optical fibre whose spiral of ≈75

(a)

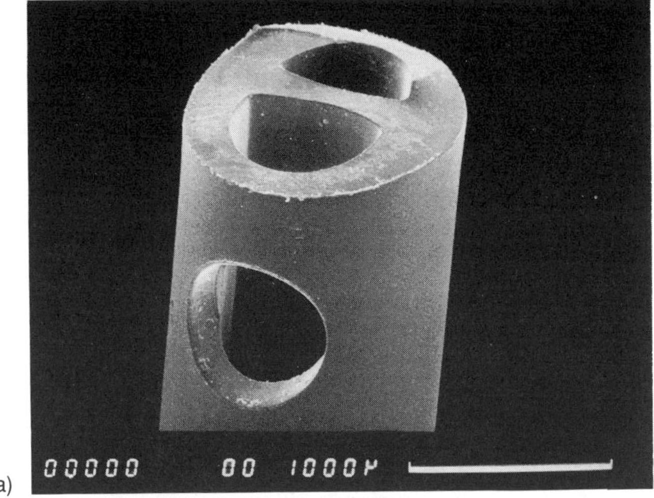

(b)

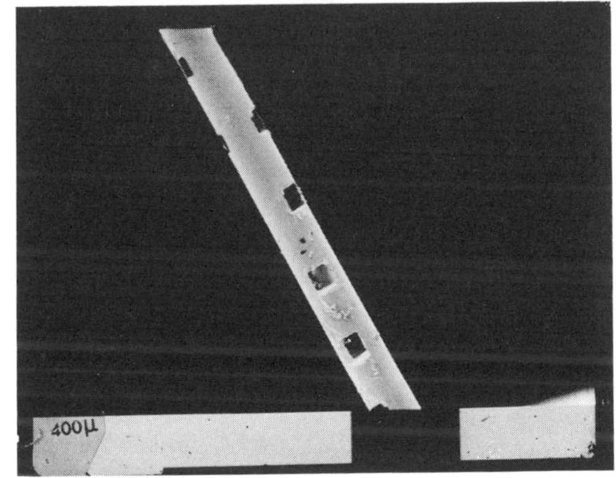

Fig. 9.20 (a) 500 μm diameter hole drilled with a KrF laser in the side of a PVC medical bilumen catheter used to measure the oxygen content in the blood of prematurely born babies, (b) a spiral of ≈75 μm holes drilled with an ArF laser in a 200 μm diameter fibre biosensor.

μm square holes is machined on a production basis using an ArF laser and a fully automated fibre feed and drilling workstation.

A further example of an excimer laser machined component is shown in Fig. 9.21. This is a mesh of ≈40 μm square hole arrays drilled into 6 μm thick PET film. After overcoating the surface with an electrically conducting layer of gold, the device is used as an accelerating/focusing grid for electrons in fast streak cameras used for measuring short-duration bursts of X-rays. The polycarbonate spinorette shown in Fig. 9.22 which is being investigated

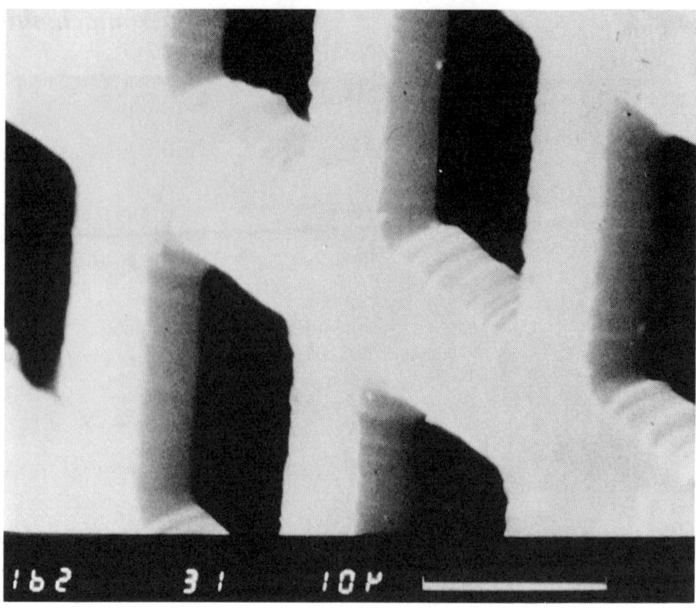

Fig.9.21 40 μm square hole meshes drilled in 6 μm thick films of PET used as accelerating and focusing grids in X-ray streak cameras.

Fig. 9.22 Triangular holes drilled with a KrF laser in a polycarbonate spinorette used for extruding synthetic fibre.

Fig. 9.23 Sinusoidal diffraction gratings in polycarbonate sheet produced by single pulse 'on-the-fly' ablation of a two-beam interference pattern created with a line-narrowed, injection-seeded KrF laser.

for use as an extrusion mould for synthetic fibres has the triangular holes in its end drilled by a KrF laser.

By using two-beam interference of a spatially (low divergence) and temporally (narrow linewidth) coherent injection-seeded KrF laser, sinusoidal gratings such as that shown in Fig. 9.23 can be ablated with a single laser pulse into materials such as polycarbonate. After overcoating the ablated area with a reflective metal such as aluminium, efficient holographic reflective gratings with up to 2400 lines/mm have been fabricated. The short-duration single laser pulse allows such fine structures to be fabricated 'on the fly' without stopping the sample during processing. If this technique can be extended to produce display holograms then it may well find widespread use in decorative and security applications that are currently common on credit cards and the banknotes of some countries. Such grating-like structures recorded with an excimer laser beam onto the cores of optical fibres are being investigated for use as input/output couplers in telecommunications applications and as reflectors in fibre optic sensors.

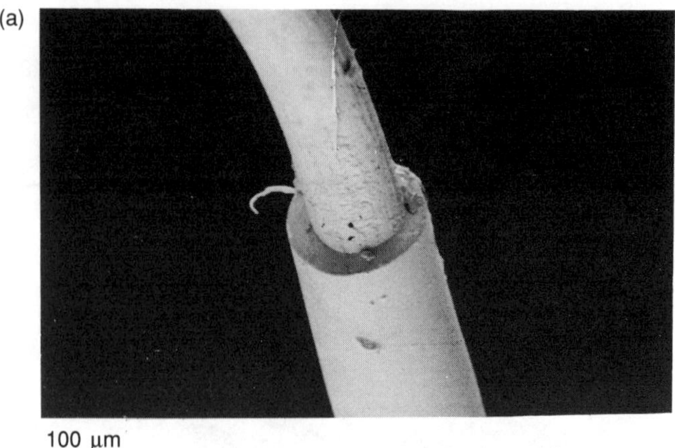

Fig. 9.24 (a) 100 μm diameter copper wire, (b) 80 μm diameter core silica fibre, whose insulation and cladding have been stripped with a KrF laser.

Excimer lasers are used on a production basis by a number of companies to cleanly strip the plastic insulation from fine wires while leaving the metallic conducting core undamaged. In Fig. 9.24 we show an example of a ≈100 μm diameter copper wire whose 20 μm thick insulation has been cleanly stripped by three-sided illumination with a KrF laser. In production the process is fully automated with computer-controlled firing of the

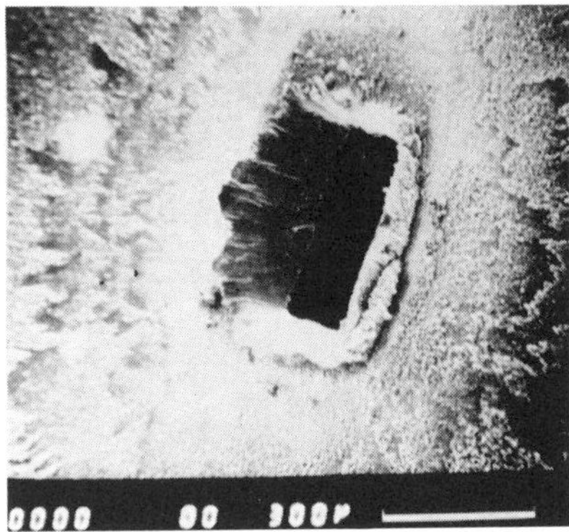
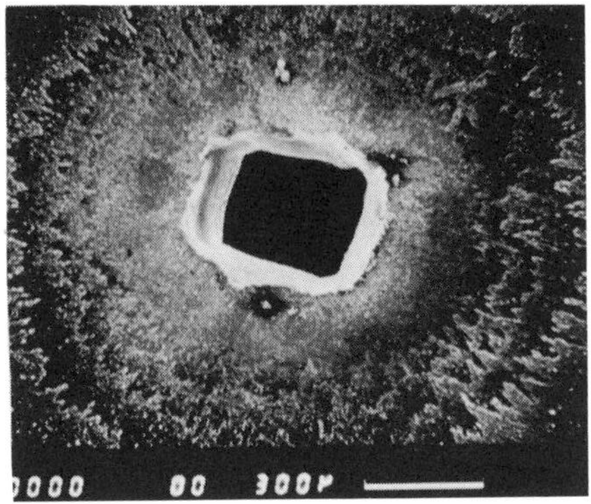

Fig. 9.25 300 μm square holes drilled in 500 μm-thick steel plate with a KrF laser at a fluence of ≈50 J/cm^2.

laser and reel-to-reel feeding of the wire. Also shown is a multimode communication optical fibre whose cladding has been cleanly removed by the laser without damaging the silica core.

Applications of excimer laser micromachining of materials are often at the forefront of industrial research and development in 'high-tech' industries and form part of sensitive key technologies that are necessarily kept confidential by the companies involved. Areas currently being researched include excimer laser machining of items such as printed circuit boards (PCBs), contact lenses, banknotes, membranes, and biological and trace impurity sensors.

9.2.2 Metals

Figure 9.25 shows a ≈300 μm square hole drilled in a sheet of steel using a focused beam from a KrF laser at 248 nm. At a focused fluence of ≈50 J/cm^2 the rate of ≈0.02 μm/pulse for drilling can be improved fivefold by flowing a jet of high-pressure helium gas into the hole as it is drilled. However initial research work has shown that for welding or cutting bulk sheet metal, excimer lasers appear to offer no significant advantages over the use of Nd:YAG and CO_2 lasers for such purposes. Perhaps when higher-power excimer lasers in the range 1–10 kW become commercially available, potential advantages such as reduced plasma screening (shorter wavelengths should penetrate further through the plasmas above the focal spot, smaller spot sizes and dross reduction will be demonstrated to be significant.

Because of the short UV photon absorption depth in metals and oxide layers, excimer lasers show great promise for modifying the surface properties of metals. For example, they can be used to clean the debris,

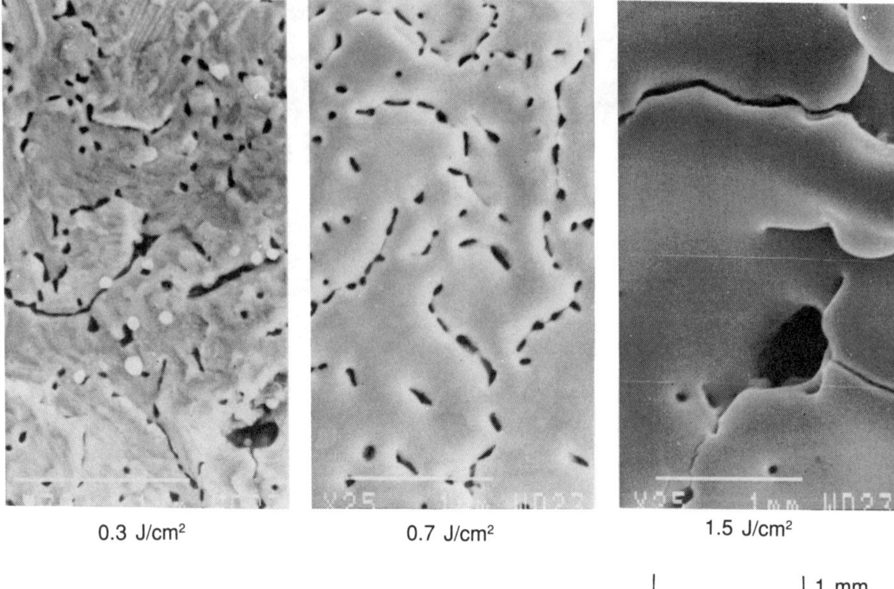

Fig. 9.26 A ≈100 μm-thick layer of boron on a nickel alloy (inconel) is melted by irradiating with a KrF laser. Its corrosion properties are increased by up to 65% following laser treatment. Photograph courtesy of Institute of Electronic Structure and Laser, Crete, Greece.

grease or oxide layers from the surface. The evaporative processes created by exposure to single-pulse fluences of a few J/cm^2 have been shown to be effective in smoothing the surface of machined metals such as steel and nickel. Such a treatment may be useful for increasing the wear-resistant properties of automotive components like camshafts, pistons and cylinder walls.

The effects of excimer laser radiation on metal surfaces are being studied as a method of improving their resistance to corrosion. Melting the top ≈1 μm of metals and alloys with a short pulse produces rapid resolidification that often leaves the material in its amorphous state. Amorphous material is homogeneous without grain boundaries which makes it much more resistant to corrosion than when in its more common crystalline state. Thus the laser can be used to glaze a protective amorphous skin on to the material. The corrosion and hardness properties of a large variety of common and exotic metals and alloys have been shown to be substantially improved using such treatment with the excimer laser. For example, research has shown that a thin layer of boron used to increase the corrosion resistance of the surface of nickel alloys can made even more resistant by melting it with an excimer laser. Figure 9.26 shows that the effect of increasing the fluence is to smooth the surface and eliminate the small-scale inhomogeneities that are present after it has been thermally deposited. The corrosion resistance has been found to be increased by up to 65% in this way. By treating the metal in the presence of a suitable reactive precursor gas (section 9.4.2), the excimer laser can also be used to plate, ion implant, dope or alloy the surface to make it harder and less susceptible to corrosion. Figure 9.27 shows a nickel alloy substrate that has had a hard boron coating deposited on to it using an ArF laser to photodissociate boron trichloride (BCl_3) gas above the surface. At higher fluences on to the surface the coating becomes more uniform.

Modifying metal surfaces by excimer laser treatment is an exciting area that shows great promise in the future for wide-ranging applications in many diverse industries.

9.2.3 Ceramics

Because they are also highly absorbing in the ultraviolet, ceramics and glasses can cleanly and precisely be etched with concentrated excimer laser light. For these harder materials the threshold fluence for etching is about an order of magnitude larger than for polymers. In Fig. 9.28, SEM pictures of structures cut into alumina, silicon nitride and zirconia are shown. The quality of the cut achieved in these tough materials is remarkable. Figure 9.29 shows the etch rate per pulse for silicon carbide cutting using three different excimer laser wavelengths. There is a distinct energy fluence threshold below which no etching takes place, while at higher fluences the process tends to saturate. As can be seen in Fig. 9.30, when working only slightly above the etching threshold, just like in polymers it is often possible to create cone- and volcanic lava-like structures. Such rough structures may be useful to enhance the adhesive bonding properties of these materials.

Since the higher fluences required for etching ceramics require that the laser beam be reduced in size by at least a factor of ten, patterning by contact replication leads to mask damage. Either focused beams or demagnified image projection of masks must be used when processing these materials. The patterns in Figs 9.28, 9.30 and 9.31 were produced by imaging.

Because of their favourable properties of hardness, chemical inertness, high-temperature thermal stability and electrical insulation, ceramic materials are in widespread use in a diverse range of industries and are difficult to machine using conventional mechanical methods. An application where excimer laser micromachining may find a role is in the fabrication of microelectronic devices where ceramic materials are often used as insulating and heat sink substrates. Feedthrough microholes, grooves, channels, localized surface roughening and other structures can readily be fabricated.

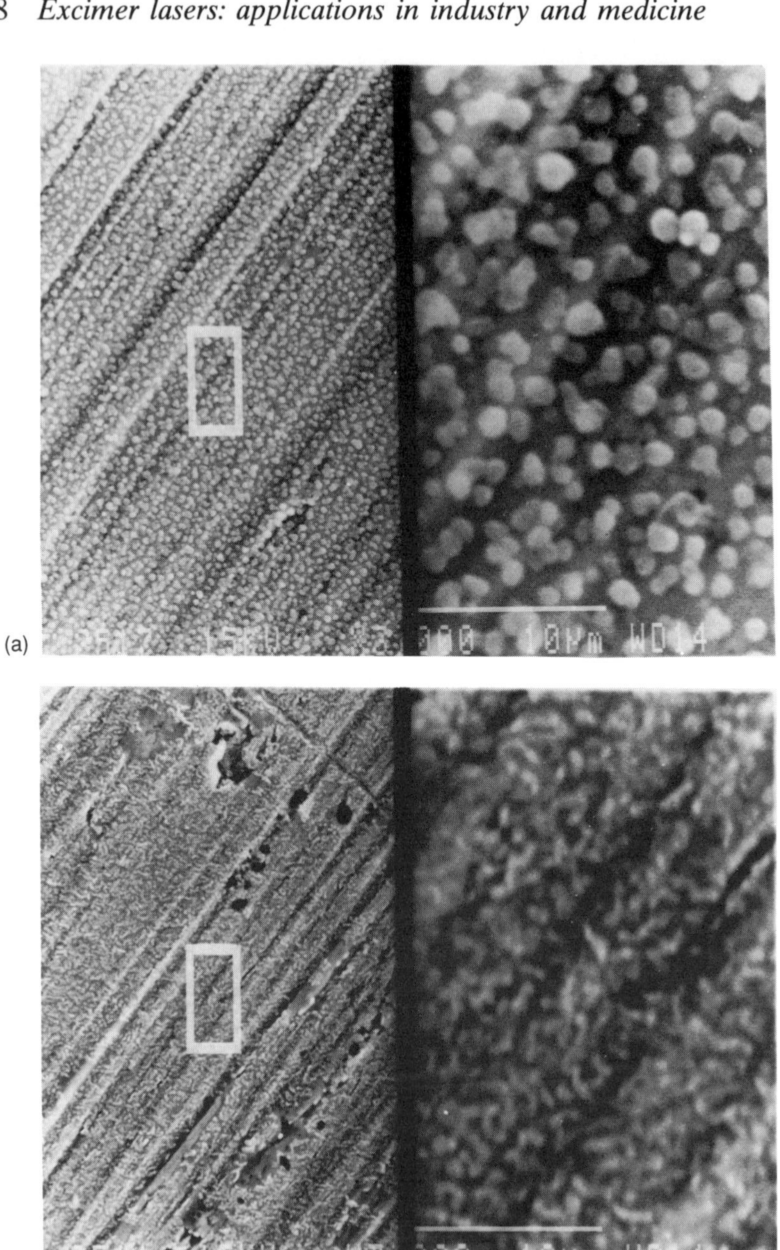

Fig. 9.27 Boron coating deposited on to an inconel substrate by LACVD from BCl_3 precursor as using an ArF laser illuminating the surface at (a) 120 mJ/cm^2, (b) 200 mJ/cm^2. Photograph courtesy of Institute of Electronic Structure and Laser, Crete, Greece.

Fig. 9.28 Structures etched in ceramics: (a) alumina with a KrF laser, (b) silicon nitride with a XeCl laser, (c) zirconia with a XeCl laser.

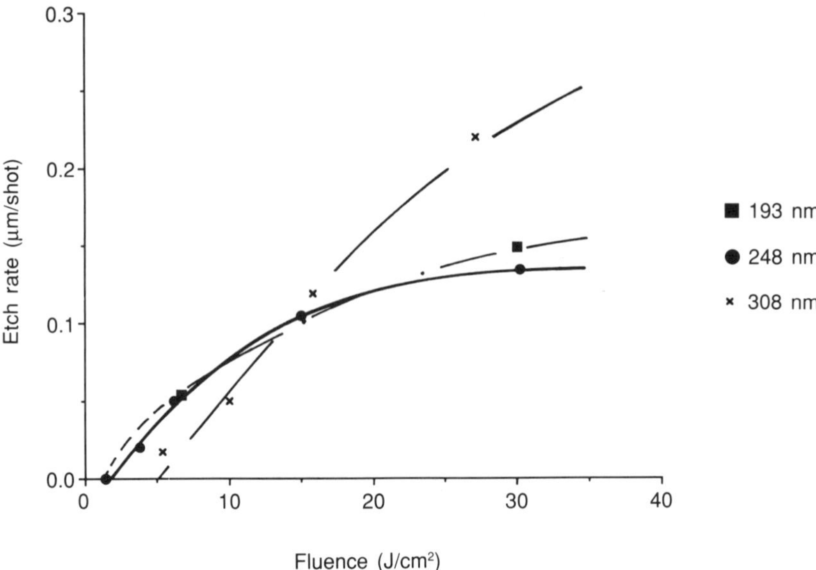

Fig. 9.29 Rate for etching silicon carbide per pulse versus fluence, with the three principal excimer lasers.

9.2.4 Glasses and crystals

Glasses are etched readily with focused excimer laser light to produce clean microcutting without damaging or shattering the surrounding material. For example, borosilicate glass is highly absorbing to light with a wavelength shorter than ≈320 nm and as shown in Fig. 9.32 can be cleanly cut at fluences similar to those required to etch metals and ceramics. The etch rate per pulse for borosilicate glass at three excimer laser wavelengths is shown in Fig. 9.33. The clean cutting provided by the laser may find applications in the specialized marking and drilling of glass-based products.

Although not a glass, synthetic fused silica SiO_2 is a common optical material. Because it is highly transmissive to UV light it is difficult to cut clean structures with excimer lasers without incurring damage to the surrounding material. An etching threshold of ≈15 J/cm^2 at 193 nm may be regarded as the threshold for causing optical damage to components such as lenses, windows and prisms that are fabricated from this material.

The relatively clean cut produced by etching a crystal of YAG – a Nd-doped 1.06 µm laser rod – with an ArF laser beam is shown in Fig. 9.34. Such a high-precision machining technique is of interest as a method for fabricating diode pumped microlasers from this material. Harder materials such as sapphire and diamond can be micromachined and drilled cleanly

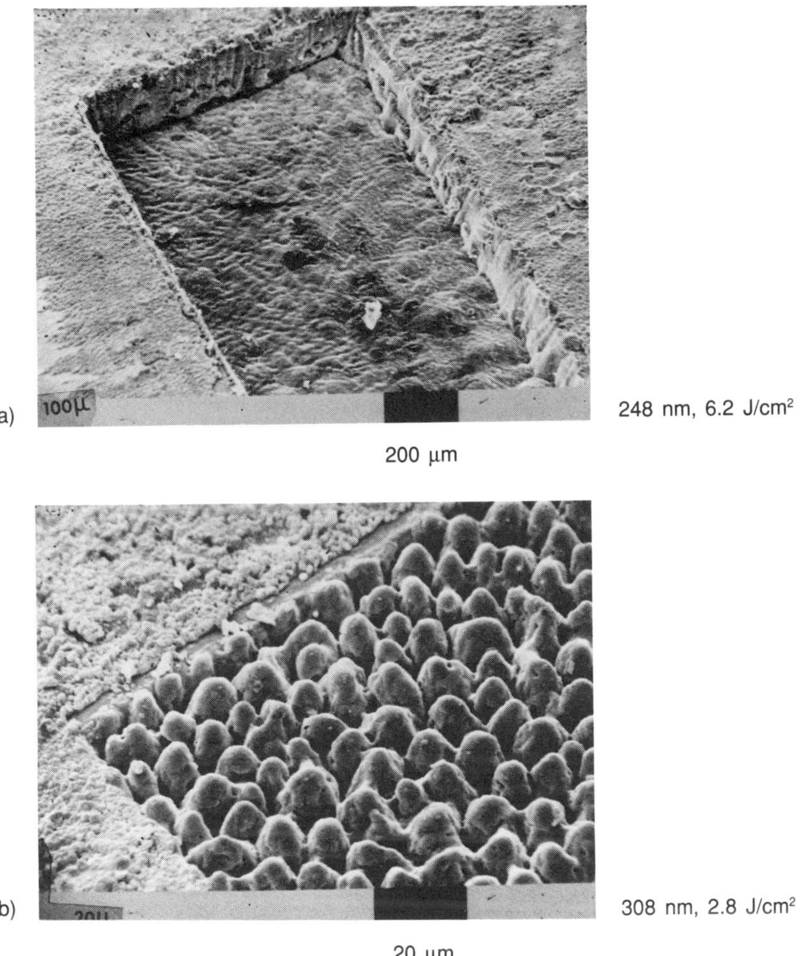

Fig. 9.30 Etches in silicon carbide obtained (a) slightly above threshold with a KrF laser, (b) at threshold with a XeCl laser.

at rates of ≈0.1 µm/pulse when using fluences of 10–100 J/cm^2 from an ArF laser.

For the high-speed GaAs integrated circuits of the future it will be important that devices are grounded to the back plane of the wafer with low capacitance connections with 'via' feedthrough holes. In 100 µm-thick GaAs wafers, via holes with diameters as small as 5 µm can be drilled in ≈1 sec with the focused output from an XeCl laser. The short 308 nm photon absorption depth of ≈13 nm in GaAs assures that there is minimal

(a)

200 μm

308 nm, 15.6 J/cm^2

(b)

40 μm

Fig. 9.31 A high-resolution patterned etched into silicon nitride using ×15 image demagnification of a mask and a XeCl laser.

lattice damage and changes in stoichiometry occurring to regions that are close to the hole.

9.2.5 Composites

Because of their extremely high mechanical strength-to-weight ratio, composites of polymers containing aramid, glass or carbon fibres are of great interest to the automobile and aerospace industries. Their toughness often means that they are difficult to machine by conventional mechanical methods. Silicon carbide or diamond tipped drills and saws wear rapidly

Fig. 9.32 A 100 µm-wide slot cut in borosilicate glass using image projection with ×20 demagnification and an ArF laser.

due to contact with the highly abrasive fibres. High-pressure water jets can be used to cut thin sheets of composites but often also have a tendency to delaminate them. While CO_2 lasers can effectively cut glass- and aramid-fibre-reinforced polymer sheets, cutting with the KrF excimer laser produces much cleaner cuts through the material and does not leave the severed fibre ends protruding from the edge. At fluences of 5–10 J/cm^2 cutting rates are ≈1 µm/pulse. While the high melting point of carbon (≈3550°C) prohibits carbon-fibre-reinforced material being cut satisfactorily with a CO_2 laser, a KrF laser readily cuts through it to leave a smooth surface without deposits and damage to the surrounding region. The porosity of the material also remains unchanged. Using a cylindrical lens to create a line focus ≈1 cm long on the workpiece in conjunction with a 100 W KrF laser source, 1 mm thick fibre-reinforced material can be cut in straight lines at rates up to ≈30 cm/min.

9.3 LASER MARKING

The marking of consumer products and components in microelectronics, medical, aerospace and automobile manufacture is an extremely important

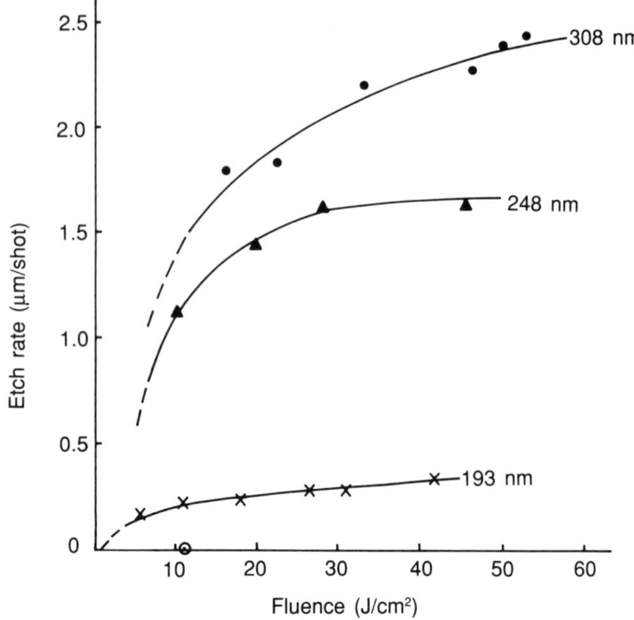

Fig. 9.33 Rate for etching borosilicate glass per pulse versus fluence, with the three principal excimer lasers.

and critical process. Depending on the type of material and the laser parameters there are several ways excimer laser radiation can leave a permanent mark on an object. Just like other pulsed lasers (principally CO_2) that are currently used to mark items from electronic components to margarine containers or soft drink cans with serial numbers or 'best before' date tags, so excimers can remove, roughen or burn material to leave an indelible mark. However, UV excimer laser photons are about ten times more expensive to produce than CO_2 laser photons, so, rather than being a more cost-effective solution, the interest in applying excimer marking to a product arises from its ability with some materials to produce a far superior mark than is possible using other lasers or methods such as dye impregnation, screen or ink jet printing. For several years in the microelectronics industry, excimer lasers have been used to print recognition characters and numbers on items such as capacitors and silicon wafers. An excimer laser wafer marking system capable of marking in excess of 100 wafers/hour is shown in Fig. 9.35. When identification marking wafers with an excimer laser, localized microcracking is less likely to occur than when using a Nd:YAG laser or a diamond scribe. Also when marking GaAs wafers the localized loss of As concentration often observed when using a YAG laser is greatly reduced with the excimer.

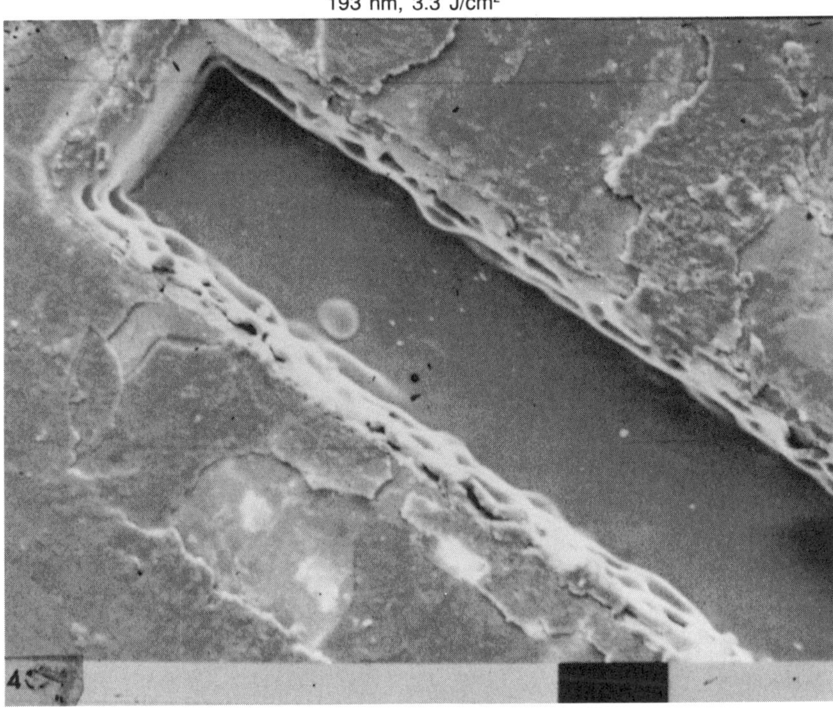

Fig. 9.34 100 µm-wide slot cut in a Nd-doped crystal of YAG using image projection with ×20 demagnification and an ArF laser.

Excimer laser marking of glass products is currently used in production for etching company logos and serial numbers on to items such as car exterior mirrors, spectacle lenses and watch bezels.

More effective use of the excimer laser to mark plastics is made by affecting a photochemical change of a pigment material loaded into the base polymer. Titanium dioxide is often incorporated in plastics, paper, paint and ceramics during their manufacture to act as a stabilizer and whitener of the bulk material. Exposure of these materials to single pulses of excimer laser light above fluences of ≈50 mJ/cm^2 photoreduces the TiO_2 pigment to leave an indelible black mark deep into the material. Choosing a base-material–laser-wavelength combination that has a relatively long photon absorption depth produces deep marking of the material and a high ablation threshold; the longer the photon absorption depth the higher the ablation threshold. Using fluences between the marking and ablation thresholds ensures blackened material is not simultaneously etched away. With 308 nm XeCl laser radiation, good single-pulse marking occurs for several per cent loading of TiO_2 into base polymers such as PTFE, FEP,

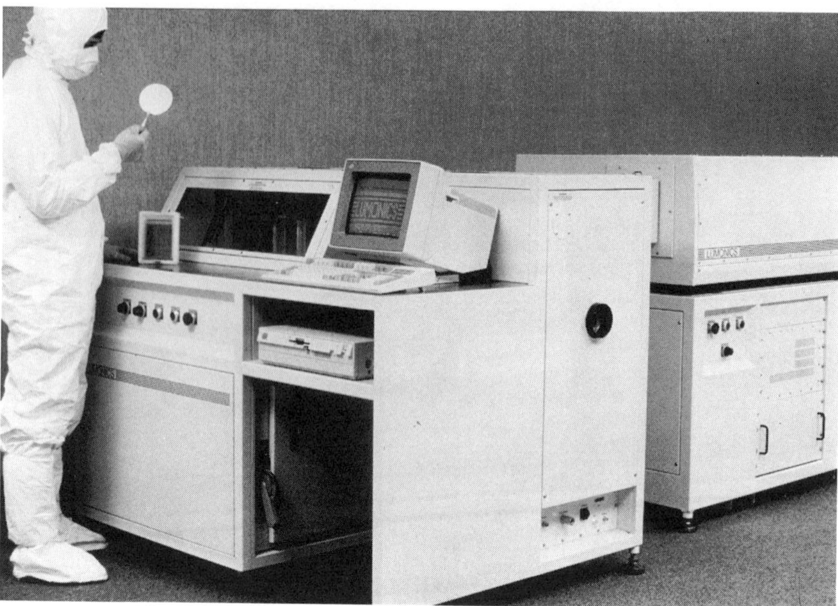

Fig. 9.35 Excimer laser silicon wafer marking system. Photograph courtesy of Lumonics Inc, Kanata, Canada.

ABS and polystyrene for which the photon absorption depth is ≈10 μm. A high-contrast mark that is extremely legible is produced without changing the basic mechanical integrity and chemical composition of the material.

Such nondestructive marking can be achieved with a single pulse and performed 'on the fly' without stopping the component. This is extremely important in critical applications such as the marking of aircraft cables where small numerals used for encoding need to be marked rapidly and should be easily legible for many years without fear of erasure by environmental or chemical attack. Several aircraft manufacturers are now evaluating and using excimer lasers for identification marking of cables for both civilian and military aircraft. In Fig. 9.36 is shown an XeCl excimer laser-marked aircraft cable whose sleeving consists of PTFE loaded with TiO_2, while Fig. 9.37 shows an excimer laser cable marking system that can mark cables with 8 or 16 character marks at rates of up to 60 m per minute.

Marks on medical components such as catheters should be highly visible and no more reactive to chemical attack than the base material. Also shown in Fig. 9.36 is a catheter used for kidney stone extraction whose length graduations are routinely marked by an excimer laser. The successful implementation of excimer laser marking of a product relies not only on

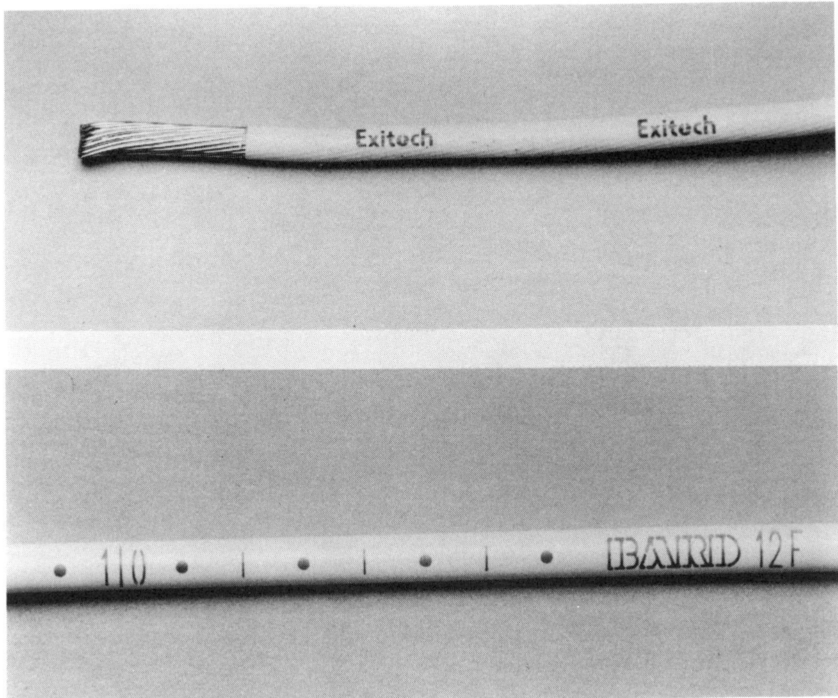

Fig. 9.36 XeCl excimer laser marked sleeving of aircraft cable and medical catheter.

the ability to produce a high-quality mark but that such a processing step be cost effective in relation to the final selling price of the component.

9.4 EXCIMER LASERS IN THE MICROELECTRONICS INDUSTRY

It has been emphasized that excimer lasers excel at microprocessing materials. The most active and demanding industry for microfabrication is in the manufacture of small electrical structures such as printed circuit boards or in microelectronic semiconductor devices in the form of integrated circuits (silicon chips). Thus it is no surprise that many research studies and feasibility trials have been carried out on the possibility of using excimer lasers for a variety of the production steps used by this industry. Because of the tight tolerances and high production volumes involved the potential application of excimer lasers to microelectronics fabrication is one of the most demanding on laser performance in terms of power, beam quality

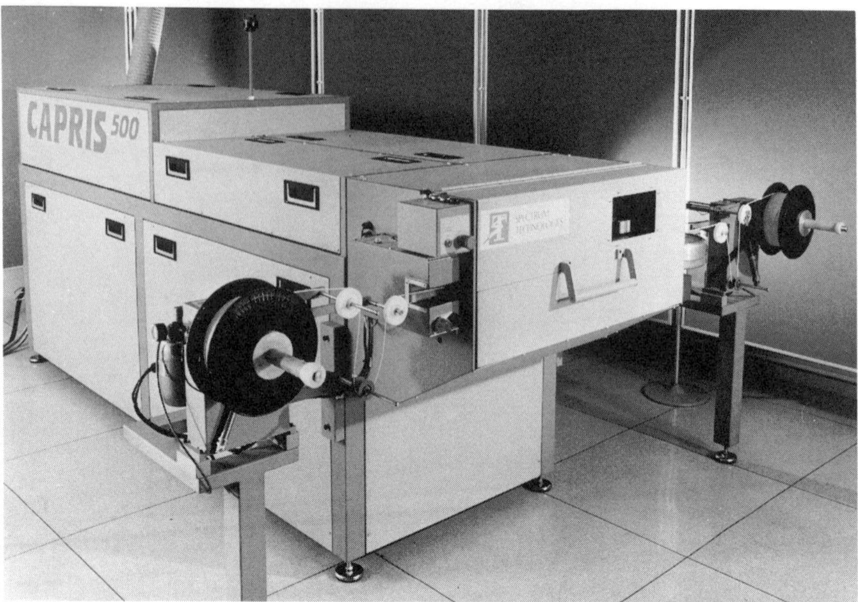

Fig. 9.37 XeCl laser aircraft cable reel-to-reel marking system. Photograph courtesy of Spectrum Technologies Ltd, UK.

and reliability. Laser manufacturers are addressing these stringent specifications by tailoring devices to suit the particular needs of the application.

9.4.1 Photolithography

Because excimer lasers are intense sources of short wavelength ultraviolet light they are of interest as source illuminators in machines that provide one of the crucial manufacturing steps required in the production of large memory size very large scale integration (VLSI) circuits. Dynamic random access memory (DRAM) devices are commonly referred to as silicon chips and a key step in their manufacture is the process known as photolithography. To understand the important potential of excimer lasers to this application it is first necessary to give a brief outline of the steps involved.

Using a turntable rotating at high speed, a thin layer of photosensitive polymer (photoresist) is spun on to a circular wafer of crystalline silicon. By previously exposing the bare silicon to steam, an electrically insulating SiO_2 layer is sandwiched between the silicon and the photoresist. As in Fig. 9.38, the photoresist is then exposed to UV light that has first passed through a mask containing details of the circuit structure to be fabricated. High-resolution lenses or mirrors are used to replicate an image of the mask on to the photoresist. Identical exposures are then either stepped

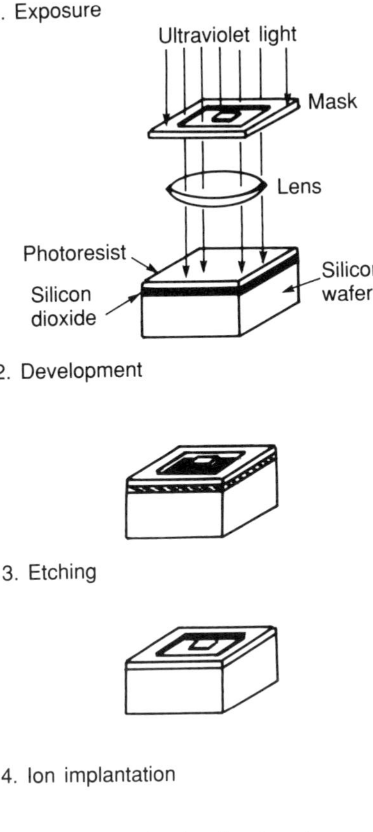

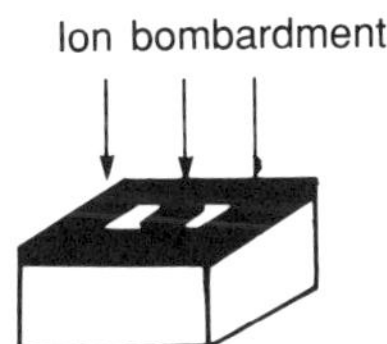

Fig. 9.38 Photolithography steps used to expose, develop and etch photoresist and ion implant silicon as used in the production of integrated circuit chips.

and repeated across the wafer or the mask/wafer combination is scanned across a slit light source. In either case, to expose the smallest sizes of circuit elements which in modern chips are ≤1 μm, the optical tolerances of the imaging optics dictate that an area of only a few square centimetres of the wafer can be exposed at any one time.

Following exposure, like in a photographic process the photoresist is chemically developed. Depending on if the resist has a positive or negative response, the exposed or unexposed regions are washed away by the

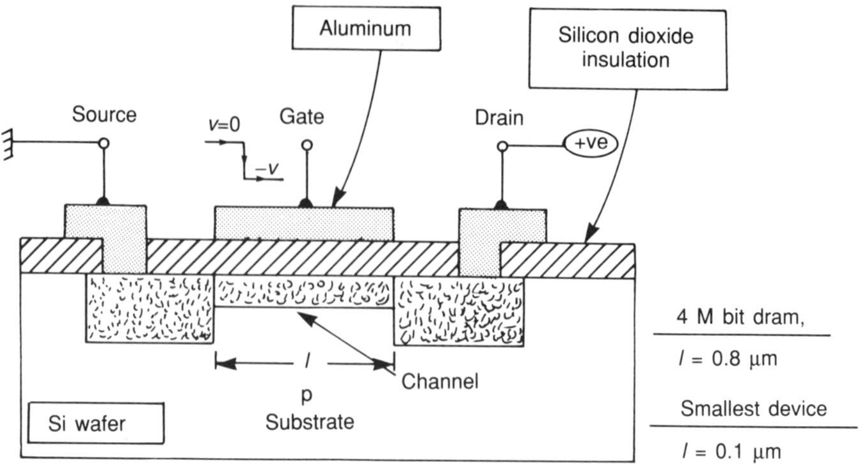

Fig. 9.39 Example of a depletion type n-channel MOSFET integrated circuit that is fabricated by several successive lithographic steps to produce the necessary sandwich-like circuit patterns of semiconducting, insulating and conducting layers.

developer. Etching the exposed silicon dioxide layer in a plasma discharge or hydrofluoric acid and stripping the remaining resist away in sulphuric acid leaves the original mask pattern replicated into the oxide layer. Depending on the type of electronic device to be produced, positive or negative ions are bombarded on to the surface which then diffuse and implant into the top layer of the bare silicon. Alternatively, an electrically conductive film of aluminium may be coated on to the oxide layer prior to the stripping of the photoresist so that sandwich-like insulating/conducting/semiconducting layers can be produced. By repeating this whole process several times, structures such as the metal oxide silicon field effect transistor (MOSFET) shown in Fig. 9.39 can be built up. The process of ion implantation tends to dislocate and damage the silicon crystal lattice at the surface, so subsequently the whole wafer is annealed in a furnace to allow it to recrystallize. Finally each wafer is diced up to produce many identical chips.

To continue to pack more devices on to the chip requires further miniaturization of the circuit elements. Higher-resolution lenses are then required for imaging ever smaller features from the mask. As mentioned in section 9.2.1, the smallest feature size, W, that can be resolved with diffraction-limited resolution in the image of an object produced by a lens or mirror is given by

$$W \approx 0.8\lambda/\mathrm{NA}$$

where $\mathrm{NA} = n\sin\theta$ is the numerical aperture of the lens/mirror – the square of which measures the light-gathering power of the system, θ is the

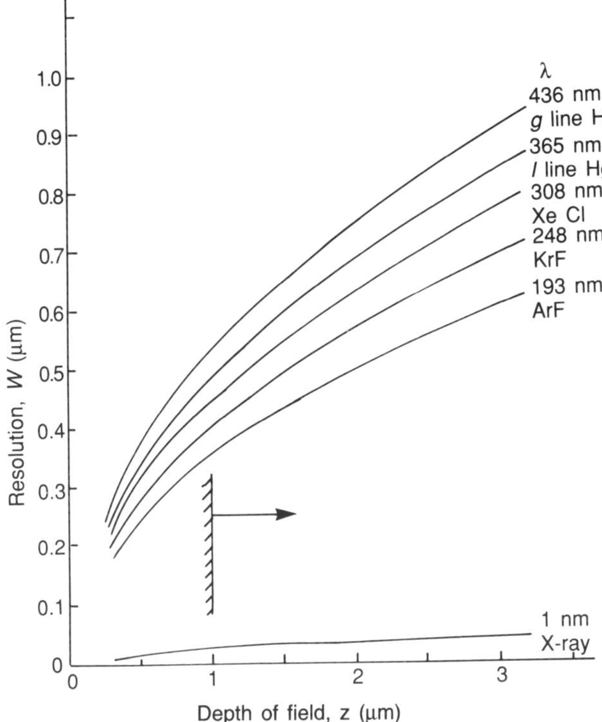

Fig. 9.40 For the most intense sources in the ultraviolet, the calculated resolution limit for imaging and contact/proximity mask replication as a function of the depth of field or when proximity processing mask–workpiece separation. For the same resolution, shorter-wavelength sources have larger depths of field. Note that with X-ray sources the highest resolution is produced even for mask–sample separations of many micrometres.

maximum light-gathering half angle of the optic and λ the wavelength of the source. Since the distance over which the smallest features are in focus is given by

$$z \approx 0.8\lambda/\mathrm{NA}^2$$

high NA lenses have short depths of focus so in practice the use of sources with shorter wavelengths to achieve higher resolution is preferable to using extremely high NA lenses. Fig. 9.40 shows the calculated image resolution as a function of the depth of field for different source wavelengths.

The current and future projections of the memory size of numbers of integrated circuit devices such as transistors, capacitors, resistors, etc. packed on to silicon chips, are shown in Fig. 9.41 in terms of design linewidth and chip area as a function of time. Currently, 4 Mbit, 1 cm^2 area chips are being produced using 6 in diameter wafers. By a combination of reducing

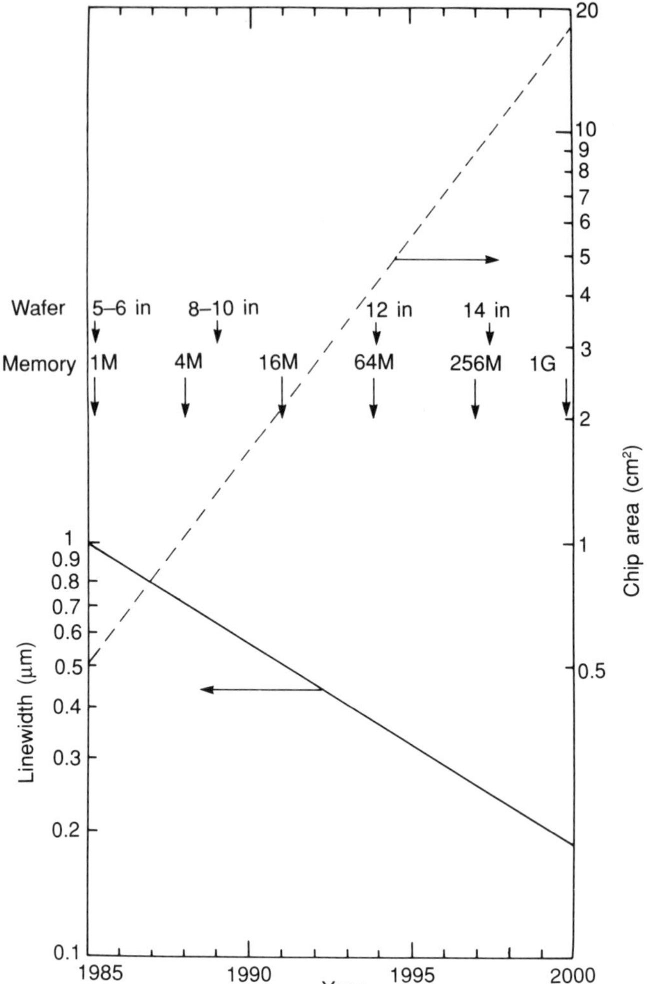

Fig. 9.41 Projections of design rule linewidths, chip area, wafer and memory size of silicon integrated circuits in the near future. 1 Gbit memory chips with 0.2 μm size circuit elements on chips having areas of up to 20 cm^2 will be produced by the end of the century.

the minimum feature size by 0.7 and doubling the chip area, during the past 25–30 years the memory size has quadrupled approximately every three years and shows every sign of continuing to do so for the foreseeable future. This remarkable progress has been achieved in part by using a combination of shorter wavelength lamp sources and higher NA lenses in the photolithography step. The highest NA lenses currently used for photolithography operate close to the limit of what is possible with modern optical design and fabrication techniques and are in the range 0.4–0.6 over flat field sizes of 15–25 mm with image demagnifications of ×5 to ×10. The source is either the *g* (436 nm) or the *i* (365 nm) lines from a powerful mercury arc lamp. The fabrication of future generations of 64 and 256 Mbit DRAM chips will require the use of even shorter wavelength sources of which there are no intense lamp sources that can compete with the average powers produced by excimer lasers. 64 Mbit chips will require feature sizes of 0.45 μm and overlay precision and dimensional control of ±0.12 μm and ±0.04 μm respectively.

Because the choice of suitable low wavefront distortion UV transmissive optical materials is limited to fused silica, MgF_2, CaF_2, and LiF, it is not possible by using materials with differing refractive indices to correct imaging lenses for chromatic aberration over the full excimer laser emission wavelength bandwidth. For example, of the above materials fused silica can be fabricated to give the lowest wavefront distortion. However, the focal length of lenses fabricated using this material alone varies by 0.03% from one side of the KrF bandwidth to the other. For a 0.4 NA lens this variation in focal length is nearly three times greater than the 1.2 μm depth of focus of the lens. The approach adopted for these systems is to use intracavity diffraction gratings, prisms or Fabry-Perot etalons to reduce the linewidth $\Delta\lambda$ of the laser emission (at present KrF, but in the future ArF) by a factor of ≥100 to $\Delta\lambda \approx$ 1–3 pm and to design the multicomponent lenses using one or two optical materials only. Since both the laser linewidth and absolute wavelength must be actively stabilized against fluctuations of ≤20%, such a bandwidth reduction involves a considerable complication to the laser hardware.

There is now a considerable drive in the semiconductor industry to develop industrialized versions of excimer lasers to act as sources with suitable image reduction lenses that can operate in the deep UV for incorporating into imaging step-and-repeat photolithographic machines. Because these KrF lasers must be able to operate reliably for three 8-hour shifts every day and produce highly stable line narrowed average powers of ≈10 W at repetition rates of up to 500 pps, this is one of the most demanding of applications for excimer lasers. Although reflective catadioptic imaging mirror systems are also capable of sub-μm imaging over large fields and have the advantage of being achromatic across the full excimer laser bandwidth, they are usually restricted to 1:1 imaging which, for the recording of sub-μm feature sizes, poses severe difficulties for mask fabrication.

Doses on the wafer must be uniform to ≤±2.5% across the exposure area. To obtain a spatially uniform illumination at the mask plane of an excimer laser step-and-repeat machine, the beam is usually split up into multiple secondary sources. This is carried out with homogenizers that incorporate devices such as fused silica diffuser plates coupled to fly's eye lenses or square kaleidoscope rods that produce several internal reflections. A KrF

Fig. 9.42 KrF excimer laser step-and-repeat system capable of imaging masks with a demagnification of 5:1 and a production resolution of 0.4 μm over a 21 mm circular field. Photograph courtesy of GCA, a unit of General Signal, Andover, MA, USA.

laser step-and-repeat machine that contains a 5:1, 0.44 NA image-reducing lens with a 21 mm diameter circular field is shown in Fig. 9.42. Examples of 0.35 μm-wide test features taken in focus and at the limits of the 1.2 μm depth of focus, as recorded in 1.0 μm-thick photoresist using a line-narrowed KrF laser source with this stepper, are shown in Fig. 9.43. The flatness of the image plane is demonstrated by the small difference in resolution of the exposure at the centre and edge of the field.

Developments are also occurring rapidly in optimizing the sensitivity and resolution of photoresists for use at 248 nm, by matching the absorption depth of the photons in the polymer with the restricted depth of focus of the lens. Contrast-enhancing and multilayer resists also show promise in achieving the ultimate performance in resolution. As shown in Fig. 9.44, one can even use a single pulse from an ArF laser to ablate and structure the photoresist layer directly. Such an ablative process may one day be used to eliminate the wet chemical development step all together and perform exposures 'on the fly'. Clearly the progress to even smaller feature sizes will continue with emphasis eventually shifting towards using a line-narrowed 193 nm ArF laser as the source. For such a source with a 0.5 NA lens and contrast-enhancing resists, the ≈0.2 μm resolution required

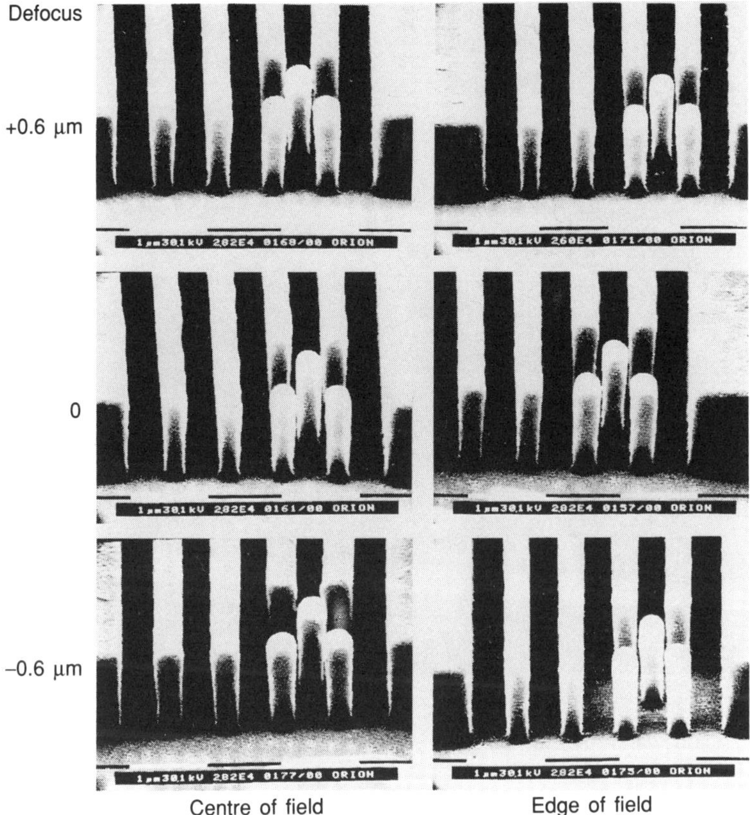

Fig. 9.43 Images taken with the stepper in Fig. 9.42 showing 0.35 μm-wide lines exposed in 1.1 μm-thick photoresist in focus and at the depth of focus. Photograph courtesy of GCA, Andover, MA, USA.

for 1 Gbit DRAM production at the turn of the century should be possible. To achieve still higher resolutions will probably require proximity mask exposure by bright sources of X-rays produced by 'table-top' superconducting synchrotrons or laser-produced plasmas.

Even here excimer lasers may play a role. It has recently been shown that by forming a plasma on copper targets in a partial vacuum or helium environment with the focused output from a high average power commercial KrF laser at an intensity of $\approx 10^{14}$W/cm^2, 1.2 keV X-rays suitable for exposing photoresist can be produced with energy conversion efficiencies of 1–10%. Thus the excimer laser could be used as a low-cost, high average power 1 nm wavelength X-ray source that exposes photoresist on wafers placed near to the plasma. This X-ray wavelength is thought to be optimum for lithography since the photons are hard enough to pass through the

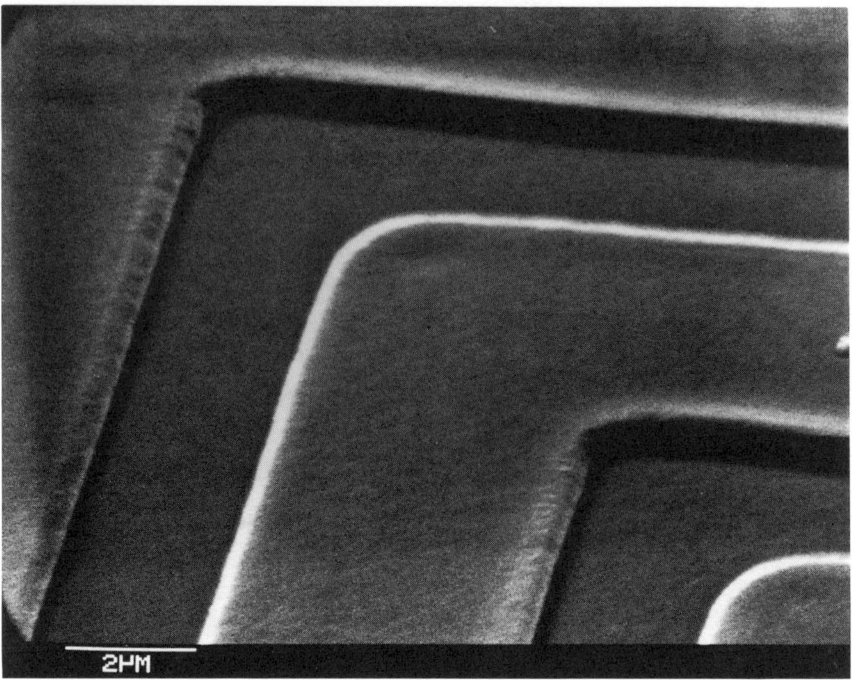

Fig. 9.44 2 μm-wide lines etched through ≈0.5 μm-thick PMMA photoresist on a silicon wafer by contact printing with a single pulse from an ArF laser at ≈0.5 J/cm^2.

boron or silicon nitride mask substrate material and not produce diffractive or blurring effects for mask–wafer proximity separations of ≈5 μm. They are also soft enough to be stopped by the gold mask circuit features and the thin layer of polymer X-ray photoresist on the wafer.

With circuit devices much smaller than ≈0.2 μm there is increasing uncertainty that silicon-based technology will be able to handle the high currents needed to drive them. To meet this challenge as well as to create devices with even greater speeds, much emphasis is now being placed on accelerating GaAs device development and its associated fabrication technologies.

9.4.2 Laser-activated chemical vapour deposition (LACVD)

The high energy of the UV photons produced by excimer lasers can be used directly to photodissociate gases which for appropriate molecules can lead to the deposition and growth of thin layers of conducting metals, semiconductors, oxide or nitride insulators. For example, by using an ArF or KrF laser to photodissociate the molecule $Al_2(CH_3)_6$ above its surface, a thin film of aluminium can be deposited on to a wafer. As mentioned in the preceding section, aluminium is extremely important as an interconnect material in the manufacture of integrated circuits. Many metals including Au, Cd, Cr, Cu, Fe, Ga, In, Mo, Ni, Pt, Sn, Tl, W and Zn have been deposited in this way. Deposition rates are usually in the thickness range 0.01–1 nm per laser pulse. When working with a 100 pps laser, rates are typically more than an order of magnitude faster than can be achieved using plasma or electron beam activated chemical vapour deposition. The great advantage of a laser-based technique for CVD is that it is a low-temperature process that can be used with temperature-sensitive substrate materials to reduce wafer warpage, dopant redistribution and defect generation. The substrate is not bombarded with energetic ions, neutrals or VUV photons and so is less invasive than plasma or electron beam assisted CVD.

In a similar manner, thin polycrystalline and epitaxial layers of pure and compound semiconductors such as Si, Ge, GaAs, InP and HgCdTe can be grown by excimer laser photodissociation of precursor gases like silane (SiH_4), germane (GeH_4) or appropriate mixtures of organometallics such as $Ga(CH_3)_3$, $Hg,Cd,Te(CH_3)_2$.

Producing insulating oxide or nitride layers on silicon wafers usually involves processing at high temperatures and often produces damage to sensitive devices like the MOSFET shown in Fig. 9.39. Machines are now available that mount vertically in a chamber two rows of wafers between which can pass axially counterpropagating beams from an ArF laser. Using the laser to dissociate precursor gases such as nitrous oxide (N_2O), ammonia (NH_3) and nitrogen trifluoride (NF_3) in the presence of silane in the chamber, high-quality insulating films of silicon dioxide or oxynitride can be deposited on to the wafers at temperatures of only $\approx 125^\circ C$. Thin insulating cyrstalline layers of alumina (sapphire), silicon nitride ceramics and diamond can be laser deposited in low-temperature reactions like:

$$Al(CH_3)_6 + N_2O + h\nu(193\ \text{nm}) \rightarrow Al_2O_3 + \text{Products}$$

$$SiH_4 + NH_3 + h\nu(193\ \text{nm}) \rightarrow Si_3N_4 + \text{Products}$$

$$C_2H_2 + H_2 + h\nu(193\ \text{nm}) \rightarrow C_x(\text{Diamond}) + \text{Products}$$

Photolytically induced dissociation of molecules can also be used to great affect by making the laser release highly reactive halogen atoms from relatively benign parent molecules that then locally chemically etch and pattern structures into the substrate.

Using the laser to heat the substrate can promote endothermic decomposition of the ambient gas at its surface and provide growth of thin films. If the gas contains halogen- or oxygen-bearing molecules then local etching or oxidation can occur. Because visible and infrared light is transmitted more readily through the gas to the substrate, pyrolytically induced CVD and etching may be more appropriate with other lasers.

However, excimer laser light provides the possibility of combining several processes together. For example, by using laser light at 193 nm directly to crack molecules such as SiH_4, BCl_3 and PCl_3 and to heat the substrate, thin epitaxial layers of silicon doped with boron or phosphorous can be grown in a single step. Using higher laser fluences to actually melt the surface of the wafer in the presence of these or other precursor gases like diborane (B_2H_6), doped surface layers that do not require subsequent annealing have been fabricated. Using these excimer laser activation methods, p-n junction solar cells having efficiencies in excess of 10% and defect-free laser diodes have been produced.

Excimer laser-assisted deposition and etching processes that can fabricate and pattern high-quality thin layers of conducting, semiconducting, insulating and doped material of relevance to both silicon- and GaAs-based integrated circuit device technologies are currently the subject of intensive research activities.

9.4.3 Excimer laser annealing

Annealing is the process used for recrystallizing the wafer following damage caused to its lattice by the ion implantation process that leaves its surface amorphous. The conventional method is to place the wafer in a hot oven for several tens of minutes. The types of device structure that can be fabricated are often limited by undesirable effects caused by prolonged high-temperature treatment of the wafer – like implanted dopant migration further into and sideways across its surface and gross wafer warpage. The use of pulsed lasers for annealing has been investigated for many years and found to offer the possibility at low temperatures of doping into the surface supersaturated concentrations of impurity ions at extremely rapid recrystallization rates through the surface layer – up to 20 m/s. Annealing with the short-wavelength excimer laser has the added advantage that melting occurs by absorption of photons more energetic than the band gap (0.72 eV in Ge, 1.12 eV in Si, 1.43 eV in GaAs). UV photons are absorbed and create electronic excitation of the semiconductor near the surface on a picosecond time scale. Several nanoseconds later the electronic states relax to excite the lattice that gives rise to heating and melting. Table 9.2 shows the absorption depths and reflectivities of some semiconductors at the principal excimer laser wavelengths.

With homogenization the excimer laser is capable of producing the highly uniform energetic beams essential for well-controlled uniform annealing. The operational range of fluence incident upon the wafer that causes melting and annealing without damage is 150–500 mJ/cm^2 at 308 nm and is appreciably wider than the window obtained with other lasers. Compared with single-pulse annealing with ruby or Nd:YAG lasers, much deeper melt front penetration can be obtained with the excimer laser without incurring damage – up to ≈1 μm at higher fluences with long-pulsed excimer lasers.

Table 9.2 Absorption depths and reflectivities of semiconductors at the principal excimer laser wavelengths

λ	351 nm		308 nm		248 nm	
	α^{-1} (nm)	*R* (%)	α^{-1} (nm)	*R* (%)	α^{-1} (nm)	*R* (%)
Si	9.3	57	6.9	59	5.5	67
GaAs	14.0	42	12.7	42	4.7	67
InP	14.5	40	14.2	38	5.6	61

Delicate high-speed silicon thin film transistors (TFTs) of the type used in shift register drivers in liquid crystal displays have been fabricated using low-temperature (≈260°C) annealing with a single pulse from an excimer laser. To form such high-channel mobility, polycrystalline films normally require annealing at oven temperatures of >1000°C, which dictates the use of expensive high melting point fused silica substrates. Low temperature annealing with an homogenized excimer laser beam will allow much cheaper glass (melting temperatures 300–600°C) substrates to be used for the displays. At ~250 mJ/cm^2 polycrystalline Si grain sizes of ~50 nm can be produced with hole mobilities in the film of more than ten times longer (up to 100 cm^2/V.sec) than in amorphous material. XeCl laser annealed p-n junction silicon photovoltaic solar cells that have efficiencies close to 20% over several square centimetres area have also been fabricated. High average power uniform beam excimer lasers offer the possibility of achieving high throughput annealing at low unit throughput costs.

9.4.4 Hole drilling, metal film removal and surface preparation

The speed at which modern computers operate is often limited by the time delays produced by the intricate interconnections between integrated circuits rather than by the chips themselves. Thus great emphasis is placed on developing efficient high-speed chip interconnection packaging. One recent method uses printed circuit boards comprising multilayer sandwiches of acrylic resin, polyimide and copper on which can be mounted and interconnected over 100 chips, depending on the number of leads to each chip. The package shown in Fig. 9.45 relies on building up, layer by layer, copper/acrylic/polyimide laminations with as many as 6000 interconnection via holes drilled between conductive metallic layers. After drilling, circuits on one layer are connected to the one below by electroless plating of copper down the via holes. Further laminations are then added on top and via hole drilling repeated. The ≈80 μm diameter via holes must be drilled cleanly with aspect ratios close to unity through each ≈75 μm-thick acrylic/polyimide laminate. The KrF excimer laser has been found to be the only satisfactory method for drilling these small holes cleanly. A conformal contact mask

Fig. 9.45 Multi-chip connection board used in a computer. Photograph courtesy of Siemens AG, Munich, Germany.

Fig. 9.46 Cross-section through a laser-drilled hole and plated micro-wiring board. Photograph courtesy of Siemens AG, Munich, Germany.

for drilling is made in the ≈18 μm-thick top copper layer by using photolithographic and wet chemical etching techniques to define the position and size of each hole. Using a laser fluence of 0.5–0.8 J/cm^2 that does not damage the copper, it takes between 300 and 400 pulses to drill simultaneously up to 60 holes through the polymer laminate. Since there is insufficient fluence to remove the copper, once through to the next layer the drilling stops automatically. Drilling the entire board involves raster scanning the beam over its surface. Figure 9.46 shows the cross-section of a 14-layer chip microwiring board made by laser drilling the via holes. After initial validation, drilling such components with 9 excimer lasers has been used successfully on the production line shown in Fig. 9.47 on a 24 hours a day, 6 days a week basis since 1988 and represents the first true use of the excimer laser on a fully automated basis. The gases in each laser are changed every 8-hour shift, its windows cleaned every week and if there is a failure it can be replaced on the line in less than an hour. The ~$30/hour operating costs of lasers in this facility are dominated by spare part replacements, in particular the high voltage thyratron switches whose lifetime is between (5–10) × 10^8 pulses.

Because most metals are highly absorbing in the ultraviolet, light from an excimer laser is absorbed in the top surface layer. As demonstrated by the copper-coated polyimide samples in Fig. 9.48 a single pulse from the laser at quite modest intensities can be used to remove and pattern, with

Fig. 9.47 Production line for excimer laser via hole drilling of laminated PCBs. Photograph courtesy of Siemens AG, Munich, Germany.

Fig. 9.48 ≈100 nm-thick copper and aluminum films on polyimide removed and patterned with a single pulse from a KrF laser at ≈100 mJ/cm^2.

high resolution, thin (<1 μm) metal films on the surfaces of many different types of substrate. This application is useful for chrome on quartz mask repair whereby the focused laser beam is used to remove erroneous metal tracks that produce faulty circuits upon replication. Faulty tracks in the circuit itself can also be with the laser. Because of the relatively low threshold fluence for metal removal in the UV compared with that for bulk material damage, the underlying surface of Si, SiO_2 or GaAs can remain undamaged after the metal layer has been removed. Automated excimer laser cutting of copper tracks on printed circuit boards as well as on polyimide tape automated bonding (TAB) ribbons that are used for interconnection and bonding of integrated circuits, are now performed on a production basis by several microelectronics companies.

When thicker than ≈1 μm, the metal film can be left intact while the laser is used to clean it or remove any thin layers of insulating polymer overcoating in preparation for solder bonding of wires. Using higher fluences of several J/cm^2, single excimer laser pulses can be used to melt metal coatings rapidly. The high surface tension and low viscosity of the molten metal creates flatter surfaces upon recrystallization. This laser planarization technique can be used to fill in small defect or contact holes and trenches that often produce bad electrical contacts. When coupling excimer laser techniques of metallic layer patterning by film removal and surface preparation to methods that also use the laser to precisely drill holes and plate surfaces, one has available an exceeding powerful 'mix and match' technology that makes mutiple uses of the laser – the potential of which is only just beginning to be realized.

9.5 FABRICATION OF THIN FILMS BY EXCIMER LASER ABLATIVE SPUTTERING

Focusing high-powered pulsed laser beams on to solid targets readily produces a plasma on the surface that not only emits short-wavelength radiation from the visible to the X-ray region, but also ablates material from the focal spot. If the target is situated in a partial vacuum then the ejected material is in the form of small particles or a vapour, which deposits a uniform thin film on most surfaces that are nearby.

Using a laser to produce evaporative reflective coatings on metals such as aluminium, silver and copper has been investigated for many years. Since the vacuum chamber requirements are modest it is an effective technique for producing metallic coatings. More conventional optical coating methods use resistive heaters, ion, electron beam or RF sputtering to remove material from the target.

The excimer laser is being used widely in research studies to produce thin films of crystalline compounds of mixed materials by ablative sputtering. Of particular interest has been the demonstration that excimer lasers can produce high-quality polycrystalline and epitaxially grown thin films of the recently discovered high-temperature ($T_c > 90K$) superconducting ceramic materials which are perovskite-like compounds of copper oxides formed from a precise stoichiometric mixture of rare or alkaline earth atoms in ratios such as: $Y_1Ba_2Cu_3O_{7-x}$, $Bi_4Sr_3Ca_3Cu_4O_x$ and $Tl_2Ba_2Ca_1Cu_2O_x$. Thin films of conventional superconducting materials have a wide range of applications in the fabrication of detectors such as superconducting quantum interference devices (SQUIDs) used to sense the extremely weak magnetic fields generated, for example, by submarine engines, geological structures and brain activity; in microelectronics for resistanceless interconnections; and in the fabrication of microwave antennae.

If a sintered pellet of the appropriate bulk superconducting material is used as a target and an excimer laser beam is focused on to its surface inside a vacuum chamber as shown in Fig. 9.49 then, for a range of angles of the substrate, it is possible to replicate in the deposited film the stoichiometry of the target compound. There is known to be at least two components that contribute to the ejected material from the focal volume. The first is a thermally evaporative component that ejects material at the various vapour pressures of the target constituents and has an approximate $\cos\theta$ angular dependence which produces the incorrect mixture on the film (θ is the angle of the ejected material from the target normal). The other is highly peaked in the forward direction – sharper than $\alpha\cos^{11}\theta$ – and is an ablatively sputtered constituent of small particles that closely replicates in the film the constituent mixture of the target. The sputtered component dominates the deposition at fluences on the pellet in excess of ≈ 1 J/cm^2 so, by positioning substrates that are parallel to the target, its stoichiometry can be replicated in the film. At quite modest focused XeCl laser fluences of several J/cm^2 with spots of a few mm size on the target, film deposition rates at ≈ 5 cm away are ≈ 0.1 nm per laser pulse and reduce as the inverse square of the distance. Thus on average each pulse deposits a film thickness of only a few atoms. With a laser operating at 50 pps, a film ≈ 1 μm thick can be grown layer by layer in about half an hour.

In Fig. 9.50 is shown a scanning electron micrograph of a patterned polycrystalline Y-Ba-Cu-O high T_c superconducting film grown by excimer laser ablation. If a hot substrate with a crystal lattice spacing similar to the film is used, then it is possible to fabricate extremely robust high-temperature superconducting films that are a single crystal (epitaxial) and do not need any thermal postannealing treatment. By comparison, other methods of depositing these films such as ion, electron beam or RF sputtering invariably produce stoichiometries in the film that are different from the target and an empirical procedure of using compensated target mixtures is usually adopted. Films produced by these methods also tend to be oxygen deficient and have to be postannealed in an oven in an O_2-rich atmosphere to make them superconducting. While molecular beam epitaxy

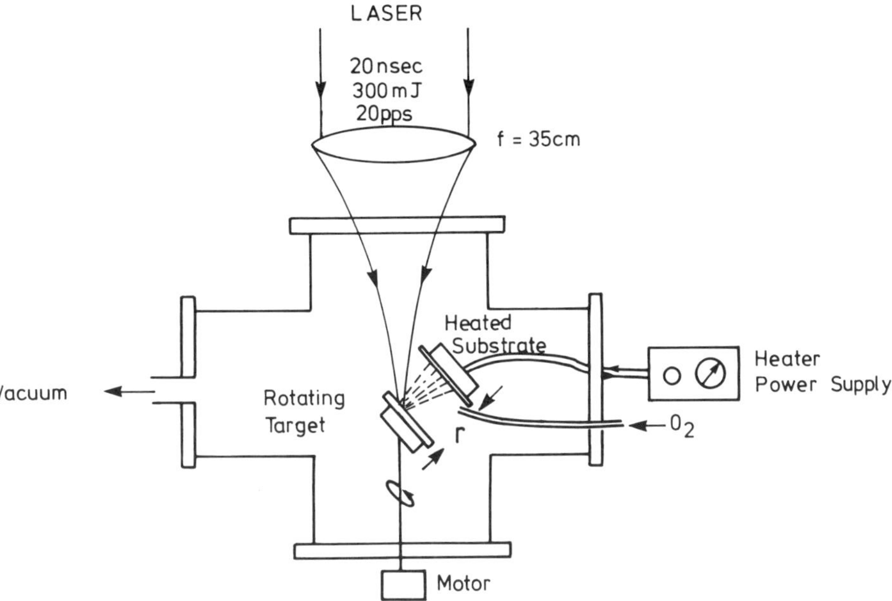

Fig. 9.49 (a) Experimental setup used to fabricate thin films of mixed compounds by excimer laser ablation of a target pellet.
(b) Excimer laser thin-film sputtering system for producing uniform top hat intensity distributions on angled targets.

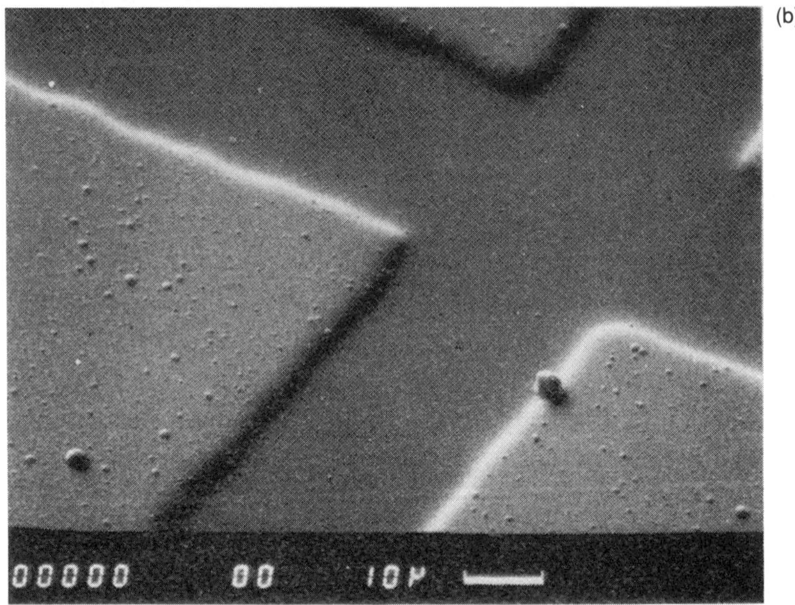

Fig. 9.50 (a) Optical and (b) electron micrographs of a YBCO film deposited through an 85 μm hole, 30 μm line mesh grid in contact with the substrate. Deposition was by ablation of a sintered pellet of the bulk superconductor at an XeCl laser fluence of ≈5 J/cm^2.

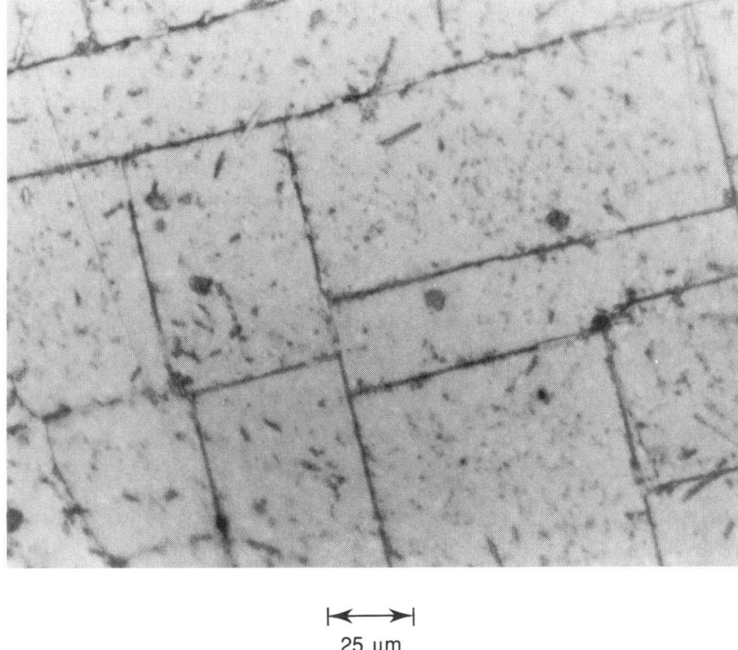

Fig. 9.51 ≈0.3 μm-thick epitaxial film of $BaTiO_3$ grown on a LiF substrate by excimer laser ablative sputtering. The mosaic structure are ferroelectric domain wall boundaries.

(MBE) can also produce good films, the equipment is extremely expensive, in excess of $1 million, and takes a day or longer to grow them. Laser ablation is similar to flash evaporation whereby cold pellets are rapidly evaporated by dropping them on to a hot surface and is a technique which also produces films having a similar mixture to the target.

For homogeneous deposition, excimer lasers have the advantage that they can produce relatively large focal spots of uniform intensity. The short-duration UV nature of the beam guarantees that the radiation is absorbed rapidly in the top layer of the target, so that heating and sputtering of material occurs in a non-equilibrium fashion without substantial phase separation occurring.

It has also been shown that excimer laser ablative sputtering can be used to grow thin epitaxial films of mixed semiconductors such as CdS as well as electro-optic, pyro and piezoelectric compound oxides such as $LiNbO_3$, $LiTaO_3$, $BaTiO_3$ and $Bi_{12}SiO_{20}$. In their thin-film form these materials have applications in microelectronics and as mechanical sensors, electro-optic waveguiding devices and infrared thermal detectors. Figure 9.51 shows an optical micrograph of a 0.3 μm epitaxial-grown film of barium titanate ablatively sputtered from a sintered pellet using an XeCl laser. The mosaic structure caused by the domain wall boundaries of this ferroelectric material are clearly apparent.

Using a fused silica microscope objective to focus an apertured low-powered excimer laser beam to a small spot, high T_c superconducting films can be serially scribed to fabricate devices such as SQUIDs. With the laser operating at higher repetition rates, and by mounting the substrate on precision stepper-motor-driven, computer-controlled x-y-z stages that keep the film at the focus of the objective, preprogrammed patterns can be scribed. The high absorption of these materials in the ultraviolet ensures that the patterning can be extremely clean and precise. The technique also works well for scribing thin films of other ceramics like $LiNbO_3$, $LiTaO_3$ and Al_2O_3, and polymer (photoresists, etc.), metal (Al, Cu, Au, etc.), semiconductor (Si, GaAs, etc.) and spin-on glass materials.

9.6 MEDICAL APPLICATIONS OF EXCIMER LASERS

Currently excimer lasers are being used to fabricate some of the small structures that form the key technology of a number of medical probes and sensors. For example, they are routinely used to drill and mark items such as the catheters shown in Figs 9.20 and 9.36 and to fabricate membranes used to detect trace elements in the blood stream.

A number of highly promising surgical procedures using the excimer laser as a scalpel are currently being evaluated. Although consisting of up to 80% water and without the single monomer unit repetition characteristic of synthetic plastics, biological tissue is also a polymer and can be cleanly etched with excimer laser light. For example it has been demonstrated that an ArF laser beam can cut a single living cell leaving an otherwise undamaged half behind. Unlike cutting tissue with CO_2 or Nd:YAG lasers, excimer laser cuts tend to be so clean that cauterising of blood vessels does not readily occur and bleeding often persists.

9.6.1 Corneal sculpting

One of the first applications of lasers to medicine was in eye surgery. For example, Nd:YAG lasers are used to remove cataracts and Ar^+ lasers to treat glaucoma and perform retinal surgery. Excimer laser light of shorter wavelength than ≈300 nm is absorbed on the front surface of the eye, the cornea, and does not penetrate through to the retina. At 193 and 248 nm the photon absorption depths in the cornea are 4 and 48 μm respectively. About 65% of the focusing power for light entering the eye is produced by refraction at the air–cornea interface with the remainder produced by the lens. For persons who are short sighted (myopic) or astigmatic there is a corrective procedure called radial keratotomy that can dispense with the need for spectacles or contact lenses. With this technique the surgeon uses a diamond knife to make a series of 4–16 radial or lateral cuts around the periphery of the cornea out of the line of vision. The ensuing relaxation of the radius of curvature weakens the refractive properties of the cornea and can compensate for the excess power contained in the lens that would normally cause an image to be focused in front of the retina. Weak myopia and astigmatism in the range –2 to –6 diopters can be semi-permanently corrected for with this technique.

Needless to say there are problems that make this procedure controversial. It is essentially irreversible and rather unpredictable in the degree of correction produced. Since the cuts must be up to a depth of ≈90 per cent of the ≈0.5 mm thick cornea, the surgeon must be extremely skilled to avoid the risk of bursting an otherwise healthy eye. There is also evidence to suggest that after a few years the cornea heals and the correction no longer remains effective. Since spectacles or contact lenses can usually provide adequate correction for the relative weak myopia that the procedure can treat, at present the relatively high risks involved for an otherwise healthy eye may not be worth taking.

It is believed that etching of corneal tissue with 193 nm ArF laser radiation occurs by photodecomposition of the peptide bonds. Since for fluences exceeding ≈0.3 J/cm^2 the removal rate of tissue saturates at ≈0.5 μm per pulse, by counting laser pulses the depth control for excimer laser cutting of the cornea is extremely precise. Figure 9.52 compares the floors of cuts made into the cornea with a diamond knife and an excimer laser. No matter how skilled the surgeon, the cuts obtained with the knife are always rough with torn tissue planes remaining. By comparison, the excimer cuts are extremely smooth with irregularities on the sub-μm scale only. The laser-treated surface also appears to be covered by a sealed pseudomembrane that prevents cells from becoming exposed.

Refractive keratectomy excimer laser procedures currently being developed pass the beam either through variable-sized or rotating apertures that are imaged with a lens on to the surface of the cornea or use a conformal sacrificial plastic mask placed in contact with it. Rather than using radial cuts, the surface of the cornea can be totally reshaped in a depth of only 50–100 μm. Depending on whether larger, smaller or parabolic surface radii are machined up to 7 diopters of myopia, astigmatism and hyperopia can be corrected (Fig. 9.53). Figure 9.54 shows front and side pictures of an eye that has had the cornea laser-machined flatter to correct for myopia, while Fig. 9.55 shows the variable-aperture circular cuts made with the excimer laser ophthalmic surgical system in Fig. 9.56. Although much of this work is still at the clinical trial stage and thus far only a few thousand human patients have been treated, the results look extremely promising and several companies in the US and Europe are marketing ArF excimer laser-based refractive keratotomy systems. One disadvantage of the technique is that cutting takes place centrally in the

Fig. 9.52 Floor of corneal keratectomy cut in the stroma of a human eye with (a) a trephine diamond knife and (b) an ArF excimer laser. Photograph courtesy of Prof. J. Marshall, Institute of Ophthalmology, University of London, UK.

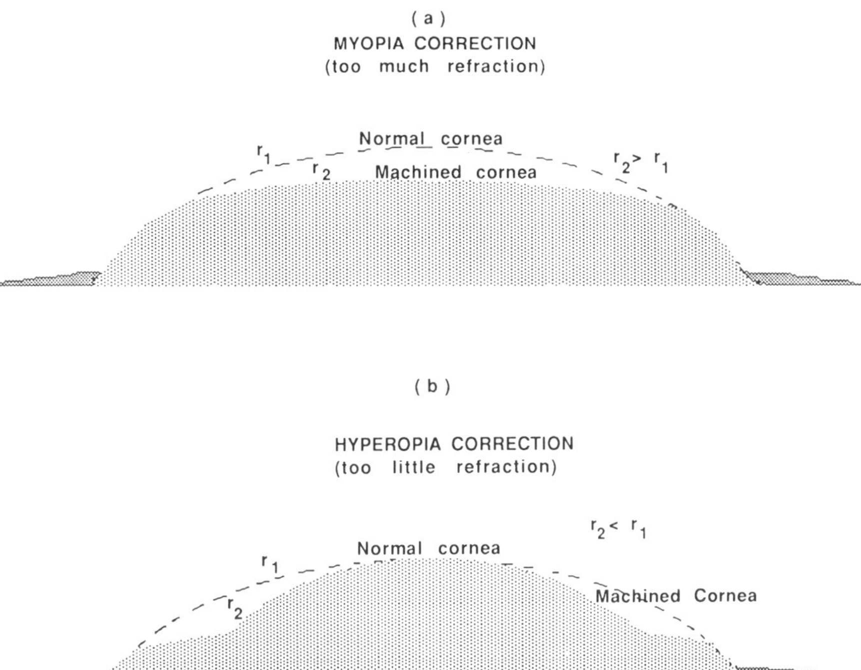

Fig. 9.53 Machining with an ArF laser and image projection on to the cornea a mask consisting of (a) a variable circular aperture producing a larger radius for myopia correction, (b) a variable annular aperture producing a smaller radius for hyperopia correction. Similar profiles can be obtained using rotating wedged slit aperture masks or sacrificial masks of variable thickness placed on to the cornea.

line of sight so that mistakes in the procedure could be potentially disastrous to the vision of the patient.

Experimental work has been performed using an ArF laser to trepan diseased corneas in preparation for transplants. Donor corneas in the form of living contact lenses which, when grafted on to the patient's own cornea, rapidly incorporate into their tissue have also been shaped with the laser. Such procedures can be used to treat diseased corneas and extreme cases of short and long sightedness.

Encouraging clinical trials have also been carried out that use a fibre-delivered excimer laser beam to cut a channel cleanly between the anterior chamber and Schlemm's canal of the eye and relieve the excess fluid pressure built up in the eyeball in cases of glaucoma.

9.6.2 Excimer laser angioplasty

The build up of calcified plaque within the arteries leads to a constriction and loss of blood to the heart that may eventually cause a heart attack

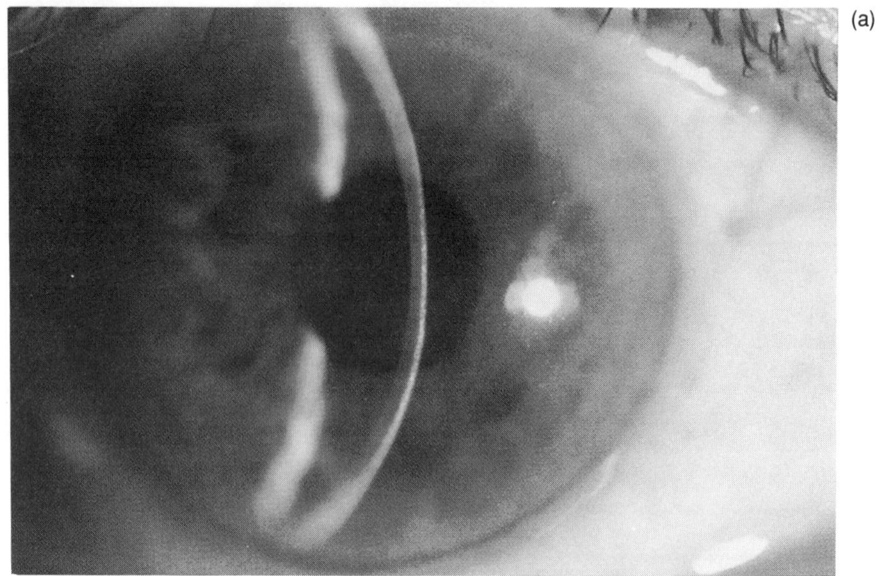
(a)

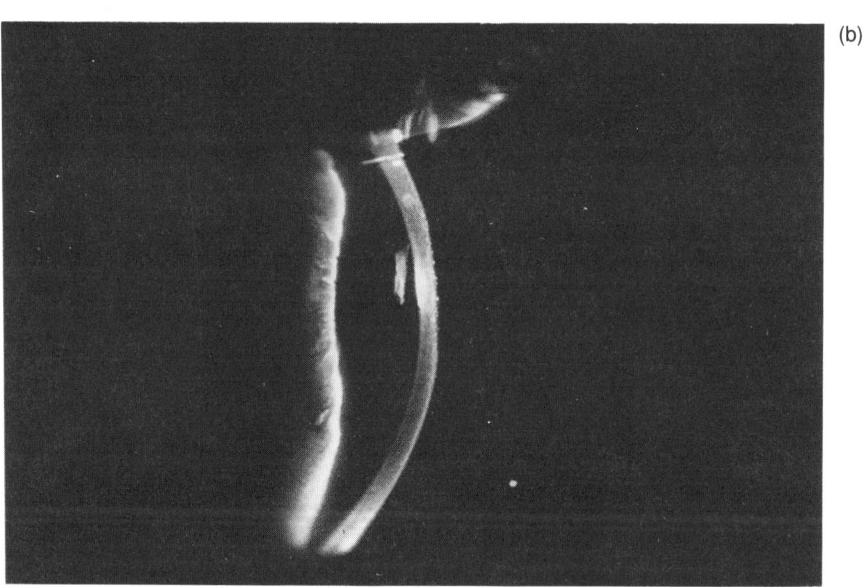
(b)

Fig. 9.54 (a) Front, (b) sideways view of an eye whose cornea has been machined flatter with an excimer laser to correct for myopia. Photograph courtesy of Taunton Technologies Inc, Monroe, Conn, USA.

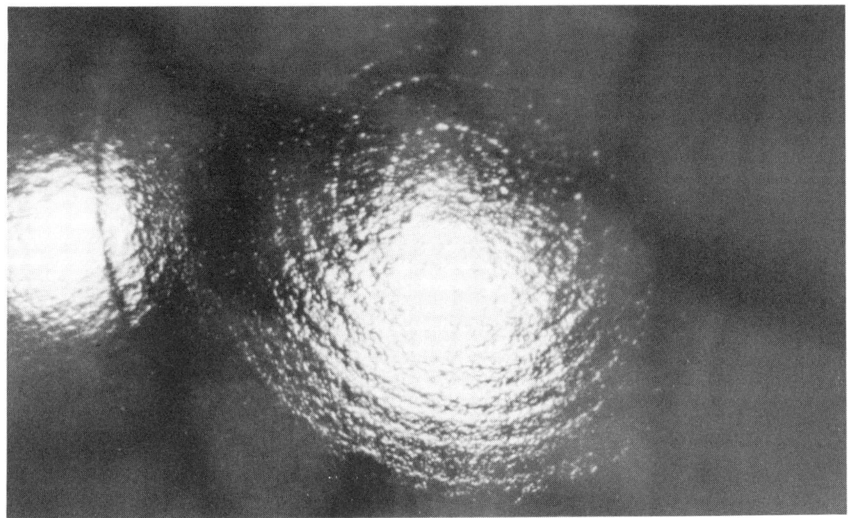

Fig. 9.55 Front view of a myopia-corrected cornea showing variable aperture cuts. The two reflected spots of light are from the ablated region (centre of photo) and the untreated region (left). Since both spots are a similar size the ablated surface is as smooth as the untreated cornea. Photograph courtesy of Taunton Technologies Inc, Monroe, Conn, USA.

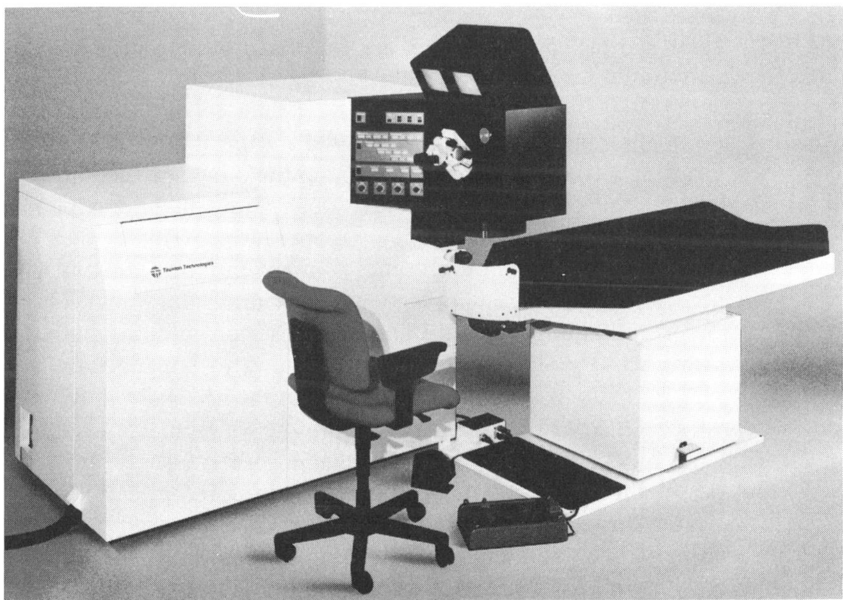

Fig. 9.56 Excimer laser ophthalmic system used for refractive surgery. The cornea is machined with 193 nm radiation at a fluence of ≈100 mJ/cm^2 in ≈30 s at a repetition rate of 10 pps. Photograph courtesy of Taunton Technologies Inc, Monroe, Conn, USA.

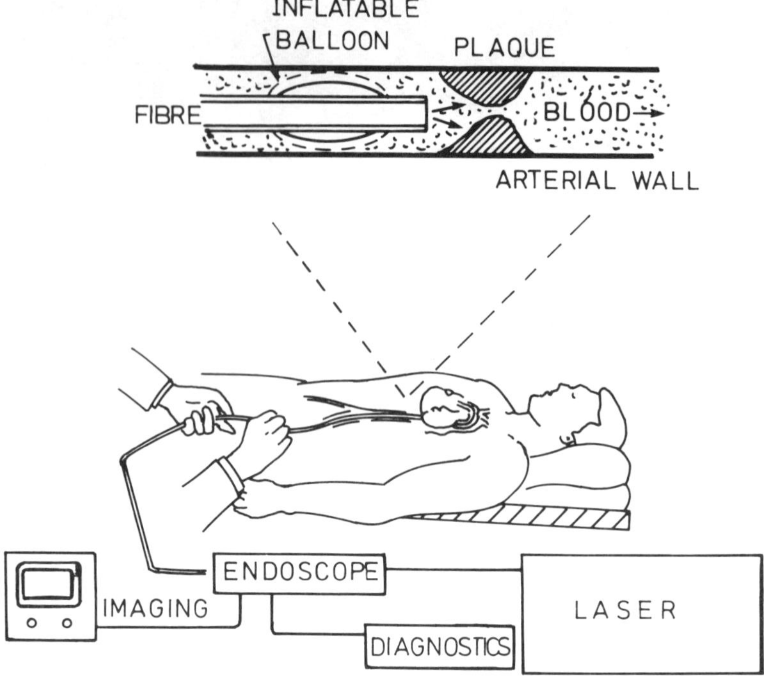

Fig. 9.57 Laser and balloon angioplasty are performed by threading the fibre catheter via arteries in the groin area to the blockage site.

which is one of the major causes of death and incapacitation in Western countries. Grafting a transplanted artery around the blockage is commonly used to alleviate the condition and every year about a million people worldwide undergo such coronary bypass operations. This is a major surgical procedure with recuperation times of many months. Less invasive is a technique called balloon angioplasty whereby under local anaesthetic a fibre catheter containing a small balloon is inserted into the patient's arterial system in the groin region. When threaded to the blockage site near the heart the balloon is inflated and stretches the arterial wall to open out the constriction. After withdrawing the fibre the wall remains stretched.

Clearing the blockage by burning through it with a laser beam (Fig. 9.57) that passes through a 200 μm–1 mm diameter core optical fibre has been investigated for several years. CW and pulsed lasers in the UV, visible and infrared have been studied – using them either to burn through the plaque directly or to heat a specially fabricated tip on the end of the fibre that then vaporizes tissue on contact. The clean cutting ability of the excimer is attractive for the direct cutting approach. Investigations have been carried out that combine the use of both the laser and balloon techniques – the laser to burn through the soft lipid-rich blockages and the balloon to open

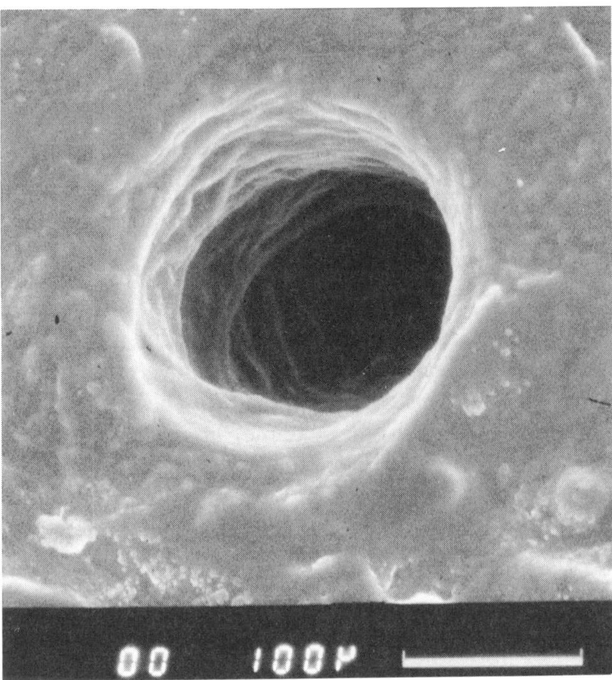

Fig. 9.58 ≈100 µm diameter hole drilled through skin tissue under saline solution using an optical fibre and a KrF laser.

out hard calcified regions that are more difficult to ablate. Other types of fibres such as lens tipped, sapphire contact and ones with slotted wall jaws have been developed that minimize the fatal risk of accidentally cutting through the healthy arterial wall. Arteries in human patients having occlusions up to 10 cm long have been successfully treated using an XeCl laser and a silica fibre to cut directly through the blockage at fluences of ≈3 J/cm^2. With the laser operating at 20 pps debulking takes 20–30 seconds to produce bore holes with diameters of up to 1 mm. An example of the cleanliness of cut in biological material that can be produced by an excimer laser is illustrated in Fig. 9.58 where an ≈100 µm diameter hole has been drilled in skin tissue with an XeCl laser beam transmitted through an optical fibre under saline solution.

Several companies now market excimer laser-based angioplasty systems for clinical trial work. Opening narrowed calcified arteries near the aortic valve, cutting of diseased coronary heart tissue, thinning down thick arterial valve leaflets and primary myocardial treatment are some of the disorders that may benefit by noninvasive percutaneous laser surgery with optical fibres. Although it has been demonstrated that excimer lasers can produce clean and precise cutting of vascular tissue through a fused silica fibre with little thermal damage, it has yet to be proven for this application that the excimer laser is superior in performance to pulsed Nd:YAG or the hot tip fibre method. Silica optical fibres are being developed that can be used to transmit efficiently high UV laser powers without incurring damage. Studies have shown that it may be possible to use the UV laser light induced fluorescence from material near the end of the fibre to aid its guiding and help discriminate between healthy and plaque tissue.

9.6.3 Excimer laser surgery

CO_2 Nd:YAG and Ar^+ lasers are now commonly used in hospitals for cutting tissue by thermal vaporization. Only recently have the cutting effects and uses of excimer lasers for surgical applications begun to be investigated. For example, it is now known that excimer laser cut skin tissue leads to a greatly reduced level of scarring upon healing compared with when the wound is made with a knife or other types of laser. The excimer laser cutting of the following types of tissue have been studied and show varying degrees of promise as possible surgical procedures:

(i) *Prostatic tissue.* 50% of all males over the age of 50 develop a prostate condition of poor and high-frequency urine flow caused by swelling and constriction of the prostate gland at the base of the bladder. A procedure that uses a fibre-delivered laser beam to cut through the blockage would be far less invasive than the current most common methods of debulking by mechanical cutting – transurethral resectioning – or balloon dilation. Such a fibre-based technique may also be useful for cutting away cancerous bladder tumours. Initial trials show that the same characteristically clean excimer laser cuts can be obtained with the prostate as with other types of tissue.

(ii) *Cancerous bile duct.* Cholangio carcinoma or cancer of the bile duct that blocks the flow of bile from the liver to the gut accounts for 1–2% of all cancers and produces distressing symptoms of jaundice and possibly death within a year. It may be possible to use fibre-delivered excimer laser light for recanalization of such blockages. Encouragingly, the data in Fig. 9.59 shows that for the same fluence of XeCl radiation the cancerous tumour cuts at roughly twice the rate as the healthy bile duct tissue.

(iii) *Neurosurgery.* The high quality and precision without incurring thermal damage is being investigated for cleanly cutting nerve fibres (axons) as applied, for example, prior to the replantation of severed limbs or to treating damaged nerves in the brain or spinal cord of people such as paraplegics. The clean preparation of the severed nerve fibre ends is of paramount importance in determining the state of recovery of the repaired

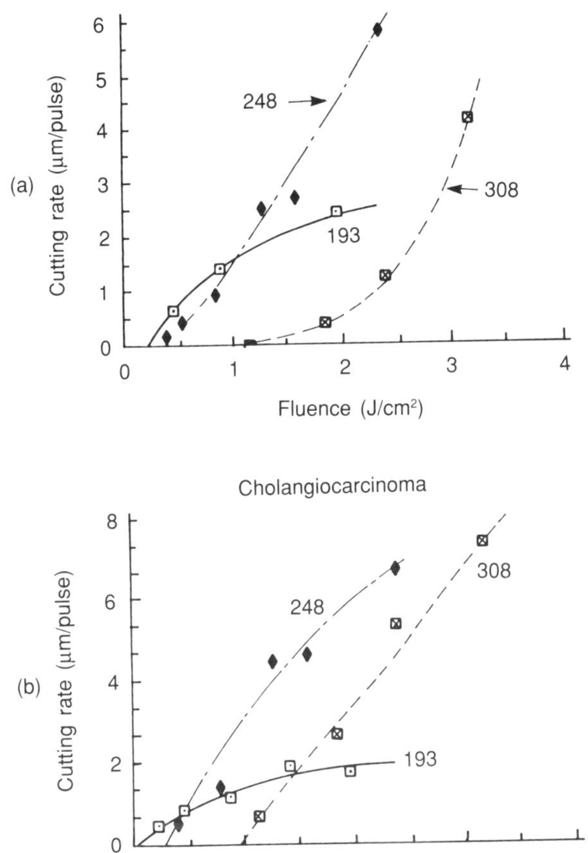

Fig. 9.59 Rate for cutting (a) healthy and (b) cancerous bile duct tumour tissue using the three principal excimer lasers.

nerve join and the corresponding usefulness of the reattached limb or spinal cord repair. Initial results with the ArF excimer laser are encouraging. It has been shown to produce cuts that are far superior to the crushed ends that tend to be produced by scissors, sharp blades or knives and the charred results with CO_2 lasers. In animal trials, the axon regeneration rate across excimer cut nerve junctions appears to be much greater than can be achieved by other cutting methods.

(iv) *Cartilage tissue.* The use of excimer lasers for removing arthritic calcified areas of cartilage while leaving behind healthy tissue in knee joints for example is being investigated. It has been shown that although the ablation rate of calcified tissue is much lower than for healthy tissue, pretreating the calcified regions in a tetracycline solution can considerably enhance it.

(v) *Kidney and gall stones.* Laser lithotripsy is a procedure whereby the fibre-delivered beam from a high-energy pulsed laser is used to smash apart kidney and gall stones in the urinary tract. The strong shock waves created by the intense laser beam as it is absorbed on the surface of the stone break it into small fragments that are washed away in the urinary tract. Experiments performed in vitreo have shown that ≈1 cm diameter gallstones can be destroyed in a few seconds with high repetition rate excimer laser pulses delivered through an optical fibre.

(vi) *Teeth.* At a fluence of several J/cm^2 an excimer laser can cut infected dentine tissue inside a tooth several times faster than the healthy dentine and leave the harder enamel on the surface untouched. As shown in Fig. 9.60 blind cuts with a KrF laser leave the underlying tissue with the same sponge-like cell porosity as the original dentine. Mechanical drills that remove the decaying region in preparation for filling of the cavity with amalgam by the dentist inevitably also destroy parts of the surrounding healthy tooth. It may be possible to use a fibre- or articulated-arm-delivered excimer laser beam to perform dental surgery on decaying teeth with less pain and destruction to healthy parts of the tooth.

Although the physiological effects of exposing living cells to intense ultraviolet light have been studied in relation to sunlight, little work has been done on assessing both the short- and long-term effects of exposure to excimer laser light at different levels and wavelengths. Exposure of the skin to modest doses of ultraviolet light is known to induce melonoma and erithema. The peak for DNA damage by single photon absorption occurs at 260 nm and one might expect the highest hazard for killing and inducing cell mutations (mutagenicity) and reducing their ability to multiply (cytotoxicity) to be for ArF and KrF lasers. There is evidence to suggest that exposure to KrF laser radiation induces the same degree of cell mutagenicity and cytotoxicity as exposure to an equivalent steady dose of radiation at the same wavelength from a lamp. Furthermore, it appears that the mutagenic and cytotoxic hazards induced by exposure to an ArF laser beam are small, perhaps because at 193 nm the photons do not penetrate deep enough into the cell to affect the nucleus. On the other hand, at longer UV wavelengths indirect damage to DNA can occur by producing oxidizing chromophores and decreasing enzyme synthesis that reduce the repair and regrowth properties of DNA. Until quantitative bio-medical and chemical data exist on the mutagenic side effects of excimer laser light on many types of tissue, some of the medical applications of these lasers may be slow in developing.

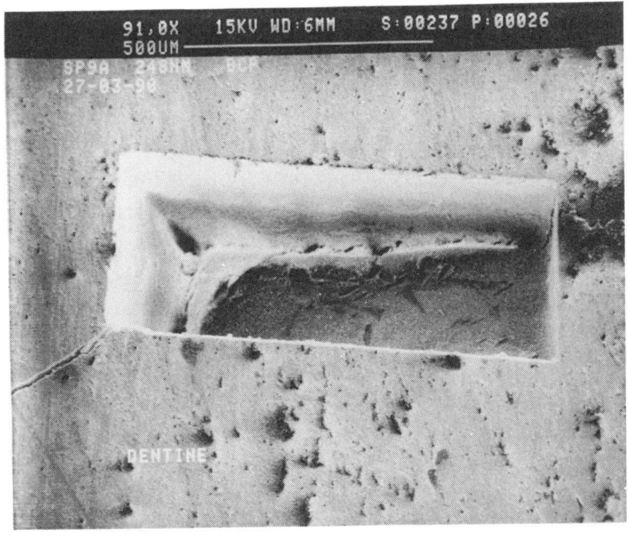

500 μm

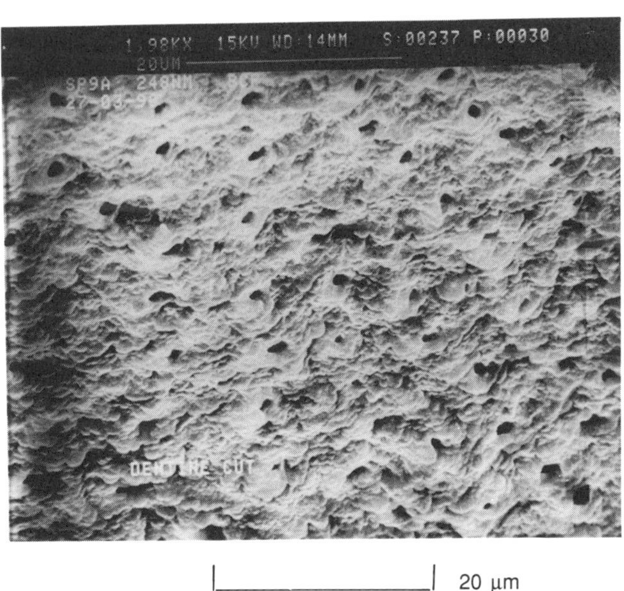

20 μm

Fig. 9.60 Slot cut in the dentine of a tooth with a KrF laser beam at ≈5 J/cm^2.

9.7 CONCLUSIONS

Some of the current and possible future important applications of excimer lasers have been introduced in this chapter and are summarized in the Appendix. Others that have not been mentioned are either commercially confidential or space limitations simply prevent their inclusion. For example, while research has shown that excimer lasers can be used for synthesising various chemicals such as ethylene and vinyl chloride, purifying gases like silane and rendering harmless various dioxine materials, it is doubtful that the processes involved would ever be economically viable when compared with more conventional methods of synthesis and treatment. The use of excimer lasers directly or to pump powerful dye lasers is also being investigated as a method for enriching isotopes – particularly those of uranium for fuelling nuclear power reactors and plutonium for military purposes. Details of this work are kept secret and since they are not truly commercial applications of the excimer laser they are beyond the scope of this chapter. Although much work needs to be done before some of the applications mentioned are realized, excimer laser development and associated applications are now two of the fastest growing areas of all laser technologies. The lasers themselves are becoming more reliable, user friendly and cheaper to buy, operate and maintain. New areas of application are continuously being created and developed.

Through collaborations such as EUREKA in Europe, AMMTRA in Japan and the SDI in the USA, the overall performance of the excimer laser will continually be improved in the next few years with high-repetition-rate multikilowatt devices becoming available.

Because of the novel way the excimer laser interacts with materials, industries that traditionally would not consider a laser appropriate for their processing needs are now often very keen to try to integrate the excimer into their manufacturing methods. Rather than replacing the more common Nd:YAG and CO_2 infrared lasers in existing applications, the excimer laser is finding its own niche of applications in a very diverse range of industries.

The potential of the excimer laser to provide the surgeon with a scalpel of hitherto unprecedented precision is obviously the main attraction of the device to applications in medicine. The possibility for safely and permanently eliminating refractive error and the need for spectacles by a 30-second treatment of the eye with an excimer laser, raises many practical, economic and even political issues whose consequences have yet to be fully discussed and even comprehended.

There is currently great optimism for future applications of excimer lasers to industry and medicine. Their applicability to a particular area and process is often not limited by the technology currently available but by the ideas and imaginations of those seeking to apply it.

ACKNOWLEDGEMENTS

The author is grateful to colleagues at Exitech, in particular Dr P.T. Rumsby, Mr C. Faulkes-Williams and Mr D. Thomas, for many useful discussions and for taking much of the cutting data presented in this chapter. He is also extremely grateful to those individuals, institutions and companies who supplied photographs of their laser treated material and equipment. The hospitality of Prof. C. Fotakis, Dr E. Honzopoulos and Dr P. Papadopoulos during writing of parts of this chapter at the Research Centre of Crete, Greece is also much appreciated.

Gathering of the original data presented in the figures, text and appendices was funded by Exitech Ltd and the UK Department of Trade and Industry, as part of the EUREKA EUROLASER EU205 Initiative.

FURTHER READING

Bachmann, F. (1989) Excimer lasers in a fabrication line for a highly integrated printed circuit board. *Chemtronics*, **4**, 149.

Bauerle, D. (1986) *Chemical Processing with Lasers*. Springer Series in Materials Science, Vol. 1, Springer-Verlag.

Boyd, I.W. (1987) *Laser Processing of Thin Films and Microstructures*. Springer Series in Materials Science, Vol. 3, Springer-Verlag.

Ehrlich, D.J. and Tsao, J.Y. (eds) (1989) *Laser Microfabrication*, Academic Press, San Diego.

Ibbs, K.G. and Osgood, R.M. (eds) (1989) *Laser Chemical Processing for Microelectronics*, Cambridge Studies in Modern Optics 7, Cambridge University Press.

Jain, K. (1990) *Excimer Laser Lithography*, SPIE Optical Engineering Press, Bellingham, Washington.

Lane, R.J. and Wynne, J.J. (1984) Medical applications of excimer lasers. *Lasers and Applications*, **59**, November.

Muller, D. (1986) Excimer Lasers in medicine. *Laser and Applications*, **85**, May.

Rothschild, M. and Ehrlich, D.J. (1988) A review of excimer laser projection lithography. *J. Vac. Sci. Techn.*, **B6**(1).

Various authors (1988) Laser interaction with tissue, *SPIE*, vol. 908.

Various authors (1988) Excimer lasers and applications. *SPIE*, vol. 1023.

Venkatesan, T. (1987) Laser deposited high T_c superconducting thin films. *Solid State Techn.*, **39**, December.

APPENDIX 9.A

Table 9.A.1 Excimer laser cutting rates of various materials. E_t, E_v and E_l are the fluences at which threshold, maximum volumetric R_v, and linear R_l, cutting rates respectively, occur

Material	λ (nm)	E_t (J/cm^2)	E_v (J/cm^2)	R_v (nl/J)	R_l (μm/shot)	R_l (J/cm^2)	Surface quality	Optimum laser
Polymers								
Polyacetylene	193	0.06	0.08	8	0.17	1.0	Smooth etch	**ArF**
Polyamide, PA	193	0.35	0.6	39	0.46	1.5	Smooth at both λs	**KrF**
(*Nylon 6 & 66*)	248	0.75	1.4	118	1.40	1.6		
Polyetherether-	193	0.25	0.30	35	0.12	0.5	Similar at all λs.	**XeCl**
ketone, PEEK	248	0.2	0.22	36	0.40	2.0	Coned at threshold,	
(*Stabar K*)	308	0.2	0.37	40	0.46	2.0	smooth above	
Polyetherimide	193	0.1	0.12	29	0.15	1.0	Similar at all λs.	**XeCl**
PEI	248	0.15	0.30	38	0.38	2.3	Coned at threshold,	
	308	0.2	0.44	35	0.62	4.2	smooth above	
Polyethersulfone	193	0.07	0.08	62	0.17	1.0	Similar at all λs.	**KrF**
PES	248	0.09	0.20	45	0.55	3.0	Smooth at	
(*Stabar S*)	308	0.2	0.74	29	1.13	5.6	0.5–1.0 J/cm^2	
Polycarbonate	193	0.1	0.18	31	0.14	1.1	Rippled surfaces,	**KrF**
(*Makrolon film*)	248	0.1	0.31	45	0.54	1.7	248 nm worse	
Polycarbonate	193	0.05	0.16	41	0.18	1.2	Smoother than	**KrF**
(*Lexan*)	248	0.12	0.32	34	0.32	2.0	Makrolon	

Polyethylene PE	193	0.25	0.5	48	0.45	1.6	Smooth etch	**ArF**
Polyethylene terephthalate PET (*Melinex*, *Mylar*)	193	0.03	0.18	56	0.16	0.3	Small surface structures	**XeCl**
	248	0.1	0.26	77	0.64	2.7		
	308	0.2	0.9	93	1.76	2.8	Poor thermal effect at 308 nm	
Polyimide, Pl (*Kapton*)	193	0.05	0.14	25	0.14	1.1	Similar at all λs.	**XeCl**
	248	0.06	0.26	38	0.44	2.8	Coned at threshold,	
	308	0.05	0.26	44	0.63	3.8	smooth above	
Poly(methyl-methacrylate), PMMA	193	0.05	0.2	37	0.44	0.8	Smooth above 0.6 J/cm^2	**ArF**
	248	0.30	1.2	310	4.2	6.0	Melting at 248 nm	
Polyparaxylylene	193	0.08	0.18	50	0.21	1.2	Smooth at both λs	**KrF**
	248	0.10	0.22	66	0.65	2.6		
Polystyrene, PS	193	0.08	0.11	38	0.13	0.5	Highly structured	**ArF**
Polyurethane PUR	248		–	–	1.0	1.5	Smooth above 0.5 J/cm^2	**KrF**
Polyphenylene sulfide, PPS	248	–	–	–	–	0.8	<0.6 J/cm^2 nodules, above rough surface	**KrF**
Polytetrafluoro ethylene PTFE (*Teflon*)	193	>2	–	–	–	–	Satisfactory etching	**F_2**
	157	0.2	0.3	50	0.3	1.2	only at 157 nm	
Polyvinylchloride PVC	193	0.05	0.22	45	0.3	0.8	Smooth etch	**ArF**
	248	0.15	0.95	36	0.65	2.4	above 0.25 J/cm^2	
Polyvinylidene chloride PVDC	193	0.08	0.26	60	0.38	1.1	Smooth above 0.15 J/cm^2	**ArF**
	248	0.40	0.93	48	0.71	2.1	Highly structured	

continued

Table 9.A.1 *(continued)*

Material	λ (nm)	E_t (J/cm^2)	E_v (J/cm^2)	R_v (nl/J)	R_l (µm/shot)	R_l (J/cm^2)	Surface quality	Optimum laser
Polyvinylalcohol PVOH	193		–	–	0.1	0.2	Structured at all fluences	**ArF**
Polyvinylidene fluoride PVDF (*Kynar*)	193	0.3	0.7	67	0.6	2.0	Highly structured	**ArF**
Nitrocellulose	193	0.02	0.6	200	1.5	3.0	Smooth etch	**ArF**
Silicone rubber	193	0.3	0.5	34	0.3	1.2	Highly structured	**ArF**
Cellulose	193	0.2	0.9	100	1.8	2.0	Rough surface	**ArF**
(*Paper*)	248	0.6	1.34	27	0.57	2.1	Damaged surface	
Cellulose acetate	248	0.46	0.71	143	1.53	1.4	Smooth	**KrF**
Composites of fibres in epoxy								
Glass fibre	308	1.0	2.5	7	2	10	Clean etching	**XeCl**
Carbon fibre	308	2.0	5	10	1.2	11		
Aramid	308	1.0	2	6	2	9		
Ceramics								
Alumina, Al_2O_3	193	1.0	5.0	0.5	0.08	32	At threshold highly	**XeCl**
	248	2.0	6.0	1.1	0.14	34	structured, appears	
	308	3.0	11.2	1.1	0.22	36	porous above	

Alumina, hipped	193	1.0	5.0	0.5	0.08	30	Smooth for all λs above 10 J/cm^2	**XeCl**
	248	2.0	6.0	1.0	0.16	40		
	308	3.0	12.4	0.9	0.23	42		
Lithium tantalate $LiTaO_3$	193	2.0	4.0	1.1	0.18	50	Smooth etch at all λs	**XeCl**
	248	1.5	4.0	1.3	0.2	45		
	308	0.8	2.5	3.6	0.35	45		
Silicon carbide	193	1.5	5	1.0	0.15	30	Cones at low fluences. Ripples at higher fluences	**XeCl**
	248	2.0	7	0.9	0.16	34		
	308	2.5	8.6	1.4	0.32	42		
Silicon carbide, toughened	193	1.5	5	1.0	0.15	34	Smooth etch at 193 nm only	**ArF**
	248	2.3	7.6	0.8	0.14	34		
	308	4.0	22.6	0.8	0.29	45		
Silicon nitride Si_3N_4, hipped	193	1.5	3	2.0	0.21	34	Cones for 248 and 308 nm below 5 J/cm^2 Ripples higher	**XeCl**
	248	2.0	5	1.7	0.20	34		
	308	2.5	4.8	3.0	0.35	40		
Silicon nitride, reaction bonded	193	0.8	3.0	2.0	0.26	38	Highly structured near threshold, rough at higher fluences	**KrF**
	248	1.5	5.2	2.2	0.27	40		
	308	2.5	7	2.1	0.36	38		
Silicon nitride, pressureless sintered	193	1.0	2.5	3.0	0.22	32	Smooth etches at all λs	**XeCl**
	248	1.5	5.2	1.6	0.21	33		
	308	2.5	6.2	2.0	0.35	42		
Silicon nitride, gas pressure sintered	193	1.0	2.0	2.2	0.22	32	Highly structured for 248 and 308 nm at low fiuences	**KrF**
	248	2.0	4.4	1.6	0.19	32		
	308	3.0	8.0	1.6	0.33	42		

continued

Table 9.A.1 *(continued)*

Material	λ (nm)	E_t (J/cm^2)	E_v (J/cm^2)	R_v (nl/J)	R_l (µm/shot)	R_l (J/cm^2)	Surface quality	Optimum laser
Silicon nitride,	193	0.8	2.0	2.5	0.2	32	Good etches >9 J/cm^2	**XeCl**
post reaction	248	2.0	4.4	1.6	0.2	34	Cones at lower	
bonded	308	3	6.2	2.1	0.36	40	fluences	
Silicon nitride,	193	0.8	2.0	2.5	0.19	30	Smoothest at low	**XeCl**
hot pressure	248	2.0	0.18	4.4	1.6	30	fluence at 193 nm	
sintered	308	3.0	5.2	2.5	0.36	44		
SiAION(α + ß)	193	0.8	2.5	1.7	0.17	32	Cones below 5 J/cm^2,	**XeCl**
	248	2.0	5	1.6	0.22	38	smooth above	
	308	2.5	6.4	1.5	0.31	42		
Lead zirconate titanate PZT	248	4.0	5	1.2	0.30	40	Smooth etch	**KrF**
$YBa_2Cu_3O_{7-x}$	193	0.23	0.5	10.0	0.17	6	Smooth etch at all λs	**KrF**
Thin film, high T_c	248	0.15	0.6	5.5	0.12	5		
superconductor	308	0.15	0.5	4.0	0.10	6		
Zirconia, ZrO_2	193	1.0	4.0	2.0	0.14	30	Smooth at all λs	**XeCl**
	248	2.0	6.2	0.9	0.15	30	above 3 J/cm^2	
	308	4.0	11.2	1.2	0.25	39		

Crystals and glasses								
Fused silica, SiO_2	193	15	–	–	0.33	30	Poor at all fluences	**ArF**
Borosilicate glass	193	4	2.5	2.4	0.32	53	Smooth above 7 J/cm^2	**ArF**
	248	8	6.2	12.0	1.68	43	Rough at 248 and	
	308	10	6.2	15.0	2.47	66	308 nm	
Yttrium aluminium garnet, YAG	193	1.5	3.0	0.89	0.06	13.5	Smooth above 3 J/cm^2	**ArF**
Silicon	193	3	4.0	1.9	0.36	15.0	Smooth	**KrF**
	248	1	1.4	6.7	0.32	12.0		
GaAs	193	0.05	3	1.4	0.26	19.0	Smooth etch at all λs	**XeCl**
	248	0.1	2	4	0.27	15.0		
	308	0.12	4	3	0.4	15.0		
Single-pulse metal film removal								
Aluminium, 0.5 μm on PET and Pl.	193	0.14	–	–	–	–	High-resolution removal	**KrF**
	248	0.32	–	–	–	–		
Copper, 0.5 μm on PET and Pl.	193	0.17	–	–	–	–		
	248	0.8	–	–	–	–		
Aluminium, 1 μm on glass.	193	1.4	–	–	–	–		
Copper, 0.5 μm on glass.	193	0.23	–	–	–	–		
	248	0.75	–	–	–	–		

continued

Table 9.A.1 *(continued)*

Material	λ (nm)	E_t (J/cm^2)	E_v (J/cm^2)	R_v (nl/J)	R_l (μm/shot)	R_l (J/cm^2)	Surface quality	Optimum laser
Biological tissue								
Cornea	193	0.1	0.2	200	0.5	0.3	Smooth cuts	**ArF**
	248	0.25	0.35	200	6.0	2.0	Some thermal damage at 248 nm	
Artery wall, *healthy*	193	0.14	0.3	100	0.6	0.9	Atheromatous tissue	**XeCl**
	248	0.25	1.0	270	1.5	1.6	cuts faster than	
	308	0.7	0.8	50	0.8	1.7	healthy aortic wall.	
atheromatous arterial wall	193	0.13	0.36	120	0.8	0.8	Smooth cutting.	
	248	0.25	1.1	420	2.2	1.5	308 nm for fibre	
	308	0.60	0.8	60	1	1.8	delivery	
Bile duct, *healthy*	193	0.3	0.7	170	2.8	3.4	Enhaced cutting	**XeCl**
	248	0.4	1.5	300	8	4.0	of cancerous tumour	
	308	0.6	3.0	400	5.5	4.0	tissue. 308 nm for fibre delivery	
cholangio carcinoma tumour	193	0.2	0.3	200	2.0	2.5		
	248	0.4	1.3	300	8.8	4.5		
	308	1.0	2.0	400	9.0	4.5		
Prostate tissue	193	0.1	0.4	100	1.5	1.5	Clean cuts at all λs	**XeCl**
	248	0.4	1.2	360	8.0	3.0		
	308	1.2	2.5	160	3.8	4.0		
Teeth *Enamel*	248	2	4	10	0.5	10.5	Clean cuts	**XeCl**
	308	1.3	–	–	0.3	2		
Dentine	248	1	4	12	0.9	18		
	308	0.3	–	–	2.0	2		

Table 9.A.2 Current and future application areas of excimer lasers to industry. The optimal laser and parameters needed are shown for each application. Medium and high uniformities refer respectively to less than about ±10% and ±5% fluctuations in fluence across the beam

Operation	Material	Application areas	Process mode	Preferred laser	Power energy	Application status
Hole drilling	Plastics Semiconductors Ceramics Metals Crystals	Via holes in PCBs. MCMs TABs and wafers. Biosensor fibres, meshes, membranes, filters, nozzles. Medical, chemical, packaging, textile components. Precious metals. Microfabrication.	Contact and projection. Med. uniformity	ArF KrF XeCl	>50 W >0.5 J	Production, R & D
Cutting	Plastics Ceramics Composites Crystals Glasses	Thin film slitting. Biomedical products. Micromachining components.	Contact and projection. Med. uniformity	ArF KrF XeCl	>50 W >0.5 J	R & D
Surface treatments	Plastics Fabrics Ceramics Glasses Metals Semiconductors	Hydrophilizing. Enhanced dyeing. Roughening. Enhanced bonding. Smoothing, alloying. Hardening, cleaning. Reducing wear	Illumination and projection. High uniformity	ArF KrF XeCl	>100 W >1 J	R & D

continued

Table 9.A.2 *(continued)*

Operation	Material	Application areas	Process mode	Preferred laser	Power energy	Application status
		and corrosion. Melting: planarization, glazing and annealing. LCVD coating. Implantation and doping				
Marking	Plastics Semiconductors Glasses Ceramics	Wire sleeving Medical devices Wafers Electronic comps: Capacitor, IC cases Glass components	Contact and projection. Uniform in 1D	KrF XeCl	>50 W >0.5 J	Production, R & D
Surface material removal	Plastics Metals Semiconductors Ceramics	Wire and PCB insulation stripping Demetallisation Mask and chip repair. Ablative direct write on resist. Scribing thin film devices. Pixellation.	Focused beam and projection. Med. uniformity	ArF KrF XeCl	<50 W <10 mJ foc. ≈0.5 J proj.	Production R & D

		Diffraction gratings } Holograms }			Line narrow – inj. seeded $\Delta\lambda \approx 3$ pm $\Delta\theta < 200$ μrad	
Lithography	Polymer resist	Lithography source for IC and mask manufacture of large memory size chips.	Projection. High uniformity	KrF	<20 W <20 mJ line narrow $\Delta\lambda \approx 1.5$ pm >50W	Preproduction, R & D
	Metal target	Production of ≈1 nm X-rays as source for lithography, 1–10 nm X-rays for microscopy	Focused beam	KrF	≈1 J low div, inj. seed $\Delta\theta < 50$ μrad	R & D
Thin film fabrication	Ceramics. Metals. Semiconductors	High T_c superconductors. Electro and nonlinear optical, piezo and pyroelectric thin films. Microlaser fabrication. Integrated optical devices	Focused or imaged beam. Med. uniformity	KrF XeCl	>50 W >0.5 J	R & D

10

Practical aspects of laser processing

P.J. Oakley

10.1 INTRODUCTION

Lasers with powers of up to 25 kW are becoming accepted increasingly as production tools for drilling, cutting, surface treatment, and welding. The highest-power machines are capable of welding and cutting thicknesses of at least 25 mm, for example, which makes the laser suitable for a wide range of applications, particularly where numbers of similar parts or assemblies are to be processed, allowing full automation and very high production rates to be achieved. Laser processes are relatively new and are still developing rapidly, one result of which is that there is a lack of appreciation of the procedures and techniques that are available now, and will be available in the future, to ensure the effective use of the processes and the quality of the processed part.

In this chapter we examine laser processing procedures and their control, and describe the techniques used to examine and assess laser processed parts in laboratory application trials and for production control. The majority of the chapter refers to welding using high-power carbon dioxide lasers, but many of the techniques are equally applicable to other processes and laser types. However, where significant different techniques are required, these are discussed in the text.

10.2 POWER MEASUREMENT

As discussed in Chapter 2, the laser output power is one of the variables in any laser processing operation. Its accurate measurement is crucial to effective process control. However, many laser users accept the power reading from the machine's calorimeter without questioning its accuracy or precision. Several aspects should be taken into account when considering

Laser Processing in Manufacturing. Edited by R.C. Crafer and P.J. Oakley.
Published in 1993 by Chapman & Hall, London. ISBN 0 412 41520 8

power measurement. Firstly, in reporting and comparing measurements, the positioning of the measuring device should be noted. In a complex laser system, with several optical components in the beam transmission system between the laser resonator and the workpiece, the laser power may be reduced by 20% or more by the time it reaches the workpiece. Ideally, power measurements for processing trials and production control should always be made at the workpiece. This is often not a practical proposition. A minimum requirement is a regularly updated factor for the power lost in beam transmission, after any measurement device, at every power level of operation.

The second point to be considered is the absolute accuracy of calorimeters and other power-measurement devices. Such devices are not usually related to any primary standards and 'kW' on one laser may be very different to those of another machine. Internally calibrated calorimeters, with integral resistance heating wires, are available which allow comparisons to be made between laser systems. These can allow some self-consistency in measurements but not absolute accuracy. The subject of power measurement is being studied currently by research laboratories and standardization organizations.

Calorimeters generally absorb the whole beam to make their measurement, and hence the power is not monitored during the actual process when the beam is directed to the workpiece. An increasing number of laser manufacturers are fitting measurement devices that sample a fraction of the beam (as described in Chapter 2), thus allowing monitoring of the laser power in-process. Whereas this is obvious preferable to knowing the power only at the beginning and end of the processing operation, care must be taken with highly reflective materials to ensure that any back-reflection of the laser beam does not interfere with the measurement.

10.3 FOCUSING OPTICS

The first decision to be made when selecting focusing optics for a carbon dioxide laser system is whether the components will be transmissive or reflective, i.e. lenses or mirrors. Above 5 kW laser power, it is normal to use reflective optics as the life of transmissive optics is unpredictable and often short at these power levels. Indeed, there are some in the laser industry who advocate the use of reflective optics on any production system, because mirrors degrade whereas lenses fail catastrophically. There are some disadvantages with mirror-based systems, however. They are usually much larger than lens systems and more costly initially, if not in running costs. Also, they cannot form the top pressure seal of a cutting gun, making them unsuitable for this process.

Mirrors for carbon dioxide lasers are made usually from gold-coated copper. The use of silicon mirrors is also quite widespread for resonator

Table 10.1 Relative merits of materials for 10.6 μm transmissive optics

	Potassium chloride KCl	Germanium Ge	Zinc selenide ZnSe
AR coatable	−	+	+
Relative material strength	−	+	+
HeNe laser wavelength transmission	+	−	+
Cost	+	−	−
Hygroscopic	−	+	+
Power-handling capability	+	−	+

optics, but less common for focusing systems. Uncoated molybdenum mirrors can be used as the final focusing optic in arduous environments, with the penalty of some power loss because of poorer reflectivity, and a higher initial cost because of the price of the material and the difficulty of working it.

Transmissive optical materials that can handle the power and the far infrared wavelength of the industrial carbon dioxide laser are rare. Three materials in common use are potassium chloride, germainium, and zinc selenide. The advantages and disadvantages of the three materials are given in Table 10.1. The ability to transmit the HeNe laser wavelength is included in the table as many laser systems use these as marker beams, and transmission through the focusing lens is therefore important. From the table it can be seen that potassium chloride is less expensive than its rivals, but is less durable. It is also hygroscopic, which means its use requires a special heated lens holder in humid conditions, and also makes it extremely difficult to anti-reflection coat reliably. It is therefore often used uncoated, with a resulting power loss of 8–10%.

The wavelength of neodymium/YAG lasers allows more conventional glass optics to be used, which have a much more highly developed design and manufacturing technology. These lasers usually have more sophisticated optical components and, as described previously, can also use fibre-optic transmission systems.

10.4 GAS-SHIELDING

The various laser processes each have different requirements of the gas shield. The simplest requirement is that for transformation hardening, where the gas shield has only to protect the beam path and focusing optic from the fumes caused by the thermal degradation of the beam absorption coating. Generally a simple co-axial flow of an inexpensive gas is used,

Fig. 10.1 Complex gas shielding shoe for laser welding.

but cross draughts of air from simple fans have also been reported as successful.

In the processes where melting takes place, the gas shield has additionally to protect the molten material from oxidation. In the laser cladding process, argon is frequently used and the composition of the gas is not considered an important process variable. However, there has been reported some effect on the process caused by different shielding gases (Jensen, 1990). As with transformation hardening, a simple coaxial shield is often used, but more complex designs may incorporate powder feeds or be in the form of reflective domes to increase the process efficiency by utilizing the energy reflected from the workpiece (Monson, 1990).

The more complex role of the gas shield in laser welding was described in Chapter 2. For the welding process, the gas shield may again range from a simple tube, often coaxial with the beam, to a complex shoe arrangement with a coaxial shield, a plasma control jet, and a trailing shield, as shown in Fig. 10.1. Although giving a much superior performance, the disadvantages of the complex shoe type of shield are its size which can restrict access, and the need to orientate it in one direction which can add complexity when welds are not linear. A general rule is that the gas shield for welding needs to be more comprehensive as the welding speed is reduced. Gas shielding is an important aspect of laser welding, which is often overlooked to the detriment of the process.

In laser cutting, the 'shielding gas' has the additional role of clearing the debris from the cut, and may also be required to provide heat by exothermic reaction with the workpiece. The gas dynamics are crucial to obtaining a tolerant process producing high quality cuts (Fieret *et al.*, 1987). This is a subject that tends to be more 'black art' than science at this stage, although research effort is currently being made to gain a more fundamental understanding of the mechanisms of the cutting process.

10.5 SEAM TRACKING AND FIT-UP FOR WELDING

If the components to be welded are made to accurate tolerances, then the workhandling used must be capable of good positional accuracy to ensure alignment of the joint line with the laser beam. For example, the fusion zone width of a butt joint in a 2 mm-thick, low-carbon steel is less than 1 mm when welded at a speed of 3 m/min and a laser power of 3 kW. In this instance, if the laser beam focus spot were positioned 0.5 mm to one side of the joint line, the fusion zone would miss the joint. The actual tolerance to joint alignment depends on the particular application, but is invariably less than ±1 mm and in general the positional accuracy of the workhandling equipment must be smaller than the minimum fusion zone width produced by the welding conditions used.

However, in large-volume production or when welding large components, the tolerance of parts cannot always match that required for consistent positioning of the joint line, so that despite the use of accurate workhandling, a missed joint will sometimes occur. Under these circumstances seam tracking is necessary to adjust the position of the joint or beam. The close-fitting butt joint found in laser welding does present a challenge to seam-tracking systems, but equipment for laser welding is available although not yet in widespread use.

Because the laser beam focused spot size is so small, close fitting butt joints are necessary if consistent quality autogenous welds are to be made. A gap of 3 mm would pose little difficulty to most arc-welding processes, but a focused laser beam would pass straight through it. The size of gap that can be tolerated depends on the thickness of the material and also whether any undercut of the weld bead can be tolerated. As an example of the importance of fit-up, good-quality butt welds without undercut cannot be made in 2 mm low-carbon steel at high speeds by conventional laser welding if there is a gap much greater than 0.1 mm between the abutting

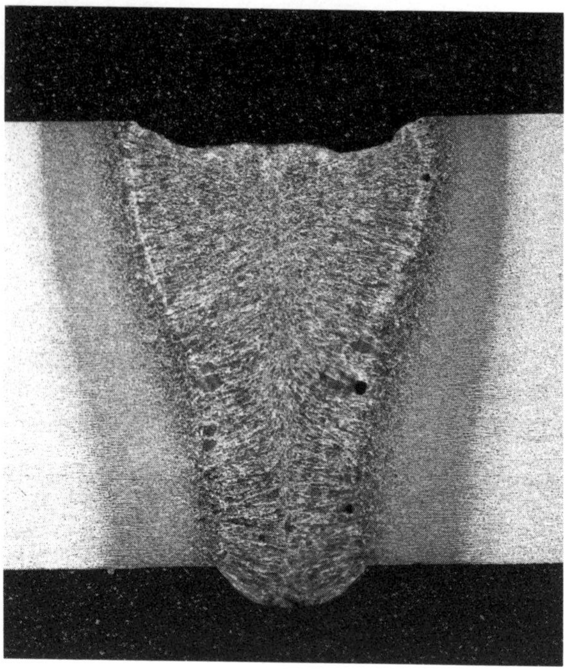

Fig. 10.2 Laser weld using wire feed to fill 2 mm gap in 8 mm-thick C–Mn steel plate. Power 5.2 kW; welding speed 0.35 m/min; original magnification of micrograph ×10.

faces. In thicker materials larger gaps can be tolerated, but gaps of more than 0.5 mm are unacceptable in all cases.

The introduction of a filler wire from an automatic wire feed system increases the tolerance to gap considerably, but does also make the process more complex. Single-pass laser welded butt joints can be made in 8 mm structural steel with a joint gap of 2 mm when using wire feed (Fig. 10.2) and it also allows thicker multipass welds to be made (Fig. 10.3). The use of wire feed for thin sheet has also been investigated, and it is rumoured to have recently been introduced into the production welding of a car body seam by a major automotive manufacturer. Another advantage of using a filler wire is that the chemistry of the weld metal can be changed to control the properties of the weld to suit subsequent processing and application requirements, such as toughness. Perhaps the ultimate system for wire-feed laser welding would sense the joint gap and send a signal to call up the relevant wire feed and laser parameters. The hardware and software for such a system is currently under development (Dilthe *et al.*, 1990).

Joint surface mismatch is not as critical as gap when laser welding. Figure

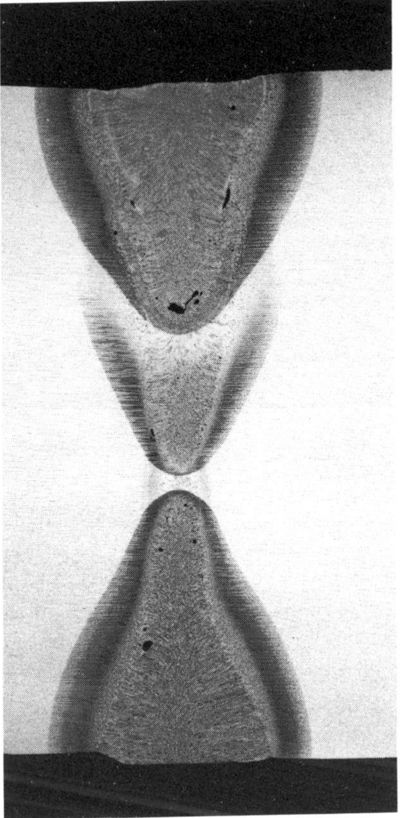

Fig. 10.3 Trial multi-pass laser weld in 25 mm C–Mn steel plate. Power 5.2 kW; welding speed 0.35 m/min for each pass; original magnification of micrograph ×5.

10.4 shows a sound weld in a butt joint made in a 4 mm low-carbon steel with a surface mismatch of 50% of the sheet thickness. A mismatch greater than this is unlikely to be tolerated in production for other reasons.

10.6 ASSESSMENT OF PROCESSED PARTS

There are no standards written specifically for the testing of laser-processed parts, so for guidance more general standards must be used. These are available from organizations such as CEN, ISO, and ASTM, covering, for example, fusion welding. Although these more generalized standards do give good background information, care must be taken as they are usually written with more traditional methods in mind and may be inappropriate for laser-processed components. The standardization agencies are beginning to recognize the need for specific standards, and it is expected

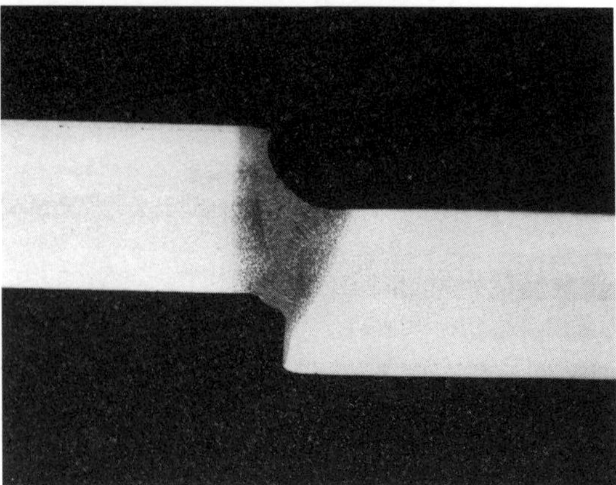

Fig. 10.4 Laser weld in 4 mm low-carbon steel sheet with a surface mismatch of 2 mm. Power 5.5 kW; welding speed 1.5 m/min; original magnification of micrograph ×5.

that these will emerge over the next few years. Meanwhile, some companies are using their own *ad hoc* standards for particular applications.

Tests used to establish the quality of laser-processed parts fall into two distinct catagories depending on the reason for making the tests. Firstly, there are tests carried out to establish the feasibility of the process for a particular application, i.e. laboratory trials or acceptance trials to determine whether or not laser processing can meet the requirements of the particular application. Clearly the tests used will depend on the application and could range from a simple visual examination to check integrity in a non-critical component, to a full evaluation of defect levels and mechanical properties of a highly critical component in, for example, an aero-engine. Secondly, there are the tests used during production to ensure that parts of the required quality are being made. Laser processes usually offer high productivity so the tests used must be quick and simple. Both the above categories of test can use destructive or non-destructive testing techniques as appropriate but, of course, destructive tests can be applied to only a small percentage of production welds. Consequently, in production, control of machine parameters is very important.

10.6.1 Non-destructive testing techniques

All of the commonly used non-destructive testing (NDT) techniques can be applied to laser welds, and the principal defects that require detection (lack of fusion, cracking, porosity) are the same as in arc welding. It is important to recognize that NDT cannot in itself tell if a weld is acceptable; all if can do is to detect defects, and it is up to the user or classification authority to decide what level of defect is unacceptable. In some mass-production industries very high levels of defects are acceptable (Eckersley, 1982).

Visual inspection, and visual inspection with aids such as dye-penetrant or magnetic particle techniques, are important particularly during laboratory trials. They are cheap to use and can be applied anywhere. However, they can detect flaws only at the surface.

Radiography is a very useful technique for detecting all kinds of defect within a weld and has been used widely on laser welds, particularly in laboratory trials. However, it requires strict safety precautions and some time to process and interpret the results, so it is not ideally suited for use on the production line. The recently developed real-time, micro-focus X-ray technique overcomes the need for time to process the results, but still has the radiation hazard.

Ultrasonic detection is the most sensitive technique and can detect defects too small to be found by other methods. However, it requires a high level of skill in interpretation and limited use is reported for laser welding. Ultrasonic detection provides data very quickly, which makes it potentially very attractive for incorporation into an in-process monitoring system. The ultrasonic process is currently being developed to operate without contacting the workpiece, using low-power laser systems, in order to monitor the laser welding process on-line. Techniques for rapid data handling are also under development for this purpose.

Laser cuts are rarely subject to sophisticated NDT techniques, as visual inspection is sufficient to identify kerf width, level of dross, and surface finish. At the other end of the scale laser transformation-hardened parts give little away under visual examination unless surface melting has occurred, and have to be sectioned for thorough testing.

10.6.2 Destructive testing techniques

One of the most widely used destructive tests is the metallographic section. A section is cut from the specimen, mounted, polished, and etched for examination under an optical microscope. A typical procedure for the preparation of a section from a laser weld has been described (*Weld. J.*, 1979). Sections of welds can be used as a simple check to determine the profile of the fusion zone (Figs 10.2, 10.3 and 10.4) or can provide detailed metallurgical data to explain the existence of defects or poor mechanical properties. For transformation-hardened or clad parts, an etched section will show the depth and morphology of the treated layer, and for cutting or drilling the thickness of the recast material and the occurrence of micro-cracks can be assessed.

Many detailed laboratory studies of laser welds have been reported in the literature for many materials, such as gold (Smith *et al.*, 1972), aluminium alloys (Moon and Metzbower, 1983), structural steels (Breinan and Banas, 1974; Willgoss *et al.*, 1979), stainless steels (Willgoss, 1979; Estes and Turner, 1974), and dissimilar metals (Seretsky and Ryba, 1976). These report data derived from sections and simple mechanical tests such as bend, tensile, and impact tests. Bend and tensile tests are relatively cheap and easy to use, so can form part of laboratory trials, or can be used in batch testing of production welds to ensure adequate strength, ductility, and freedom from defects.

When carbon steels are welded the laser weld can be much harder and stronger than the parent material because of the rapid cooling rate experienced by the weld. This can make interpretation of these tests in laser-welded carbon steel more difficult than normal, as deformation and failure almost always occur in the parent material and not in the weld, but even so a useful check of weld quality is obtained.

As section thickness increases, brittle fracture is a greater risk and so impact properties become important. The results of Charpy and dynamic tear tests in laboratory trials are reported in the literature (Willgoss, 1979; Stoop and Metzbower, 1978). Good properties in Charpy and dynamic tear testing are reported but, as has been pointed out by Goldak and Nguyen (1977), these tests may be misleading in cases where the crack path does not follow the weld but diverts into the parent material. As the weld is much stronger than the parent material this can happen even though the weld is more brittle than the parent material. Figure 10.5 shows a Charpy test specimen where the fracture path has run into the parent material. As Fig. 10.6 shows, it is possible to draw a Charpy curve through the points representing only specimens which failed in the fusion zone. This seems a valid approach as the results are true weld values and are the worst case at any temperature.

However, for welds in higher-strength steels, all tests at ductile failure temperatures may deviate. Crack tip opening displacement (CTOD) testing, which evaluates the initiation of fracture rather than initiation and propagation, may give more meaningful results. CTOD testing is unfortunately very expensive and to date its use on laser welds is not widely reported. The use of modified Charpy specimens to overcome the problem of deviation has been investigated (Hall and Wallach, 1989), but none has yet been adopted as a standard technique.

The properties of welds under cyclic loading, i.e. in fatigue, are particularly important in some applications but little work appears to have been done on the fatigue properties of laser welds, although some work has been reported on low-alloy steels and titanium alloys (Fraser and Metzbower, 1983), and ship-building steels (Brook, 1988).

For surface-treated specimens, hardness measurement is often the most important routine test, but more sophisticated techniques, such as wear or corrosion testing, may be necessary, depending on application.

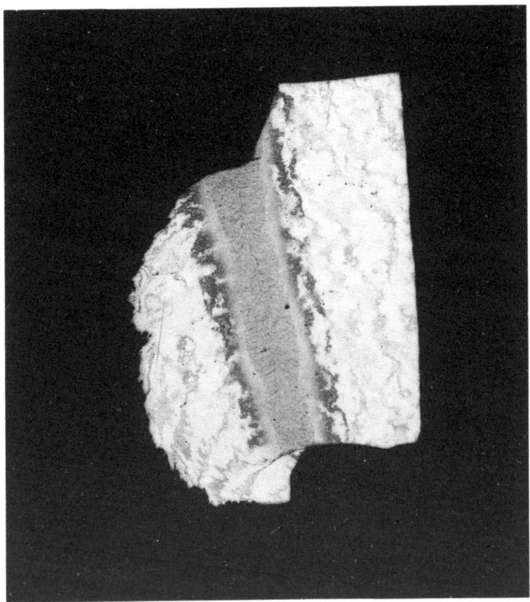

Fig. 10.5 Charpy test specimen, showning the fracture path deviating from the laser weld in the parent material.

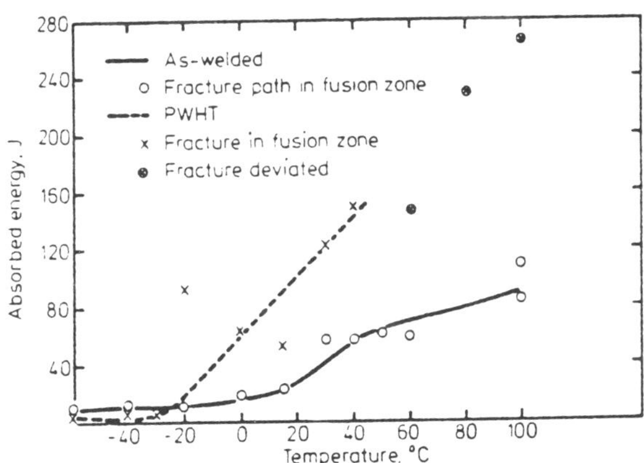

Fig. 10.6 Charpy impact test results for laser welds in 12.5 mm-thick, BS 4360, Grade 50D structural steel. Power 6 kW, welding speed 0.36 m/min.

REFERENCES

Anon. (1979) Checking the quality of laser welds. *Weld. J.,* **58**(7), July, 53–4.

Breinan, E.M. and Banas, C.M. (1974) *Preliminary Evaluation of Laser Welding of X-80 Artic Pipeline Steel*, WRC Bulletin 201, December, 47–57.

Brook, S.J. (1988) Laser skid welding of T-joints for ship fabrication. *Proc. 5th Int. Conf. on Lasers in Manufacturing* (LIM5). IFS Publications, September, 165–75.

Dilthey, U. *et al.* (1990) Laser beam welding with wire feed – Management of variable gaps and influence on weld pool metallurgy. *TWI Power Beams Technology Conf.*, Stratford, Sept., 69–78.

Eckersley, J.S. (1982) CO_2 laser welding of aluminium air spacers for insulated windows. *Proc. Conf. ICALEO*, Boston, September.

Estes, C.L. and Turner, P.W. (1974) Laser welding of a simulated nuclear reactor fuel assembly. *Weld. J.*, **53**(2), February, 66s–73s.

Fieret, J. *et al.* (1987) Overview of flow dynamics in gas-assisted laser cutting. *Proc. SPIE Conf. on High Power Lasers*. SPIE vol. 801, April, The Hague, 243–50.

Fraser, F.W. and Metzbower, E.A. (1983) Solidification structure and fatigue crack propagation in laser welds. *Proc. Conf. Applications of Lasers in Material Processing II*, January, Los Angeles.

Goldak, J.A. and Nguyen, D.A. (1977) A fundamental difficulty in Charpy V-notch testing narrow zones in welds. *Weld. J.*, **56**(4), April, 119s–125s.

Hall, B.E. and Wallach, E.R. (1989) Microstructure and properties of autogenous and wirefeed laser welds in steel plate. *Proc. Conf. Advances in Cutting and Joining Processes*, TWI, Harrogate.

Jensen, T.A. (1990) *Dynamic Powder Feeding System for Laser Cladding*, Report 90.39, Danish Welding Institute (Svejscentralen), DK-2600 Glostrup, Denmark.

Monson, P.J.E. (1990) Comparison of laser hardfacing with conventional processes *Surface Engineering*, **6**(3), 185–93.

Moon, D.W. and Metzbower, E.A. (1983) *Laser beam welding of aluminium alloy 5456. Weld. J.*, **62**(2), February, 53s–58s.

Seretsky, J. and Ryba, E.R. (1976) Laser welding of dissimilar metals: titanium to nickel. *Weld. J.*, **55**(7), July, 208s–211s.

Smith, D.L. *et al.* (1972) Laser welding of gold alloys. *J. Dental Research,* **51**(1), 161–7.

Stoop, J. and Metzbower, E.A. (1978) A metallurgical characterisation of HY-130 steel welds. *Weld. J.*, **57**(11), November, 345s–353s.

Willgoss, R.A. *et al.* (1979) Assessing the laser for power plant welding. *Weld Metal Fab*, **47**(2), March, 117–27.

Index